TAKING COMPLEXITY SERIOUSLY:

Policy Analysis, Triangulation and Sustainable Development

TAKING COMPLEXITY SERIOUSLY:

Policy Analysis, Triangulation and Sustainable Development

by

Emery Roe
University of California, Berkeley, U.S.A.

KLUWER ACADEMIC PUBLISHERS
Boston / Dordrecht / London

Distributors for North America:
Kluwer Academic Publishers
101 Philip Drive
Assinippi Park
Norwell, Massachusetts 02061 USA

Distributors for all other countries:
Kluwer Academic Publishers Group
Distribution Centre
Post Office Box 322
3300 AH Dordrecht, THE NETHERLANDS

Library of Congress Cataloging-in-Publication Data

A C.I.P. Catalogue record for this book is available
from the Library of Congress.

Printed on acid-free paper.

Printed in the United States of America

Dedication

Especially for Ken Hall and Andrés Davidovits

Contents

ACKNOWLEDGMENTS

It is no secret that you can't be a practicing policy analyst and be definitive at the same time. Policy analysts never have the last word, they could always use more information or time, and this is nowhere truer than when dealing with complex and uncertain policy issues. Yet there is little doubt analysts specifically and students of policy generally could be doing a better job in addressing such issues.

First, consider the sorry state of the policy analysis profession. Turn to the *Journal of Policy Analysis and Management*, and thumb through the pages devoted to "Curriculum and Case Notes" over the last decade or so. Two things will strike you: how extravagant the space devoted to the teaching of public management,[1] how meager the ink spilt for analysis. Far too little attention has been given to what is, for practicing policy analysts such as myself, the more urgent matter: how better to analyze—not just manage—policy issues of high uncertainty and complexity. You'll not find one item in "Curriculum and Case Notes" since its inception in 1986 to the time I started writing this book (end of 1994) that deals directly with this need to teach innovative approaches to policy *analysis*.[2] When I tried to correct that deficiency by writing the first version of what is now Part I of *Taking Complexity Seriously*, my contribution was returned because, well, the journal only publishes pieces on public management there. Go figure.

This book also takes up the unfinished business of the last one. *Narrative Policy Analysis* showed how contemporary literary theory could be used to

---

[1] Nor does a quick browse indicate any real improvement in the 1995 and 1996 *JPAM* volumes. See whole or in part: Behn (1986), Diver (1986), Elmore (1986); Robyn (1987); Lynn, Jr, (1987), Bardach (1987), Behn (1987); Elmore (1989), Heifetz et al (1989); Leone (1989); Behn and Brough (1990); Barzelay and Kaboolian (1990); O'Hare (1991); Brock (1991); Arnold and Leone (1992); Averch and Dluhy (1992); Brock (1993); Brock (1993a); Lynn and Jiping (1994); and Brock (1994). It appears that the more recent editors of "Curriculum and Case Notes" even see little or no difference between analysis and management!

[2] Some of these concerns are touched upon in O'Hare (1991, 523). Brock (1991, 722), Elmore (1986), Bardach (1987, 196). The conventional policy analytical answer to uncertainty and complexity—reduce 'em—can be found in Doron (1986) and Quigley and Scotchmer (1989).

analyze and make recommendations about policy scenarios developed under conditions of extreme uncertainty, complexity and polarization. By the time that book was finished, it had become clear to me that several other "unconventional" (i.e., non-probabilistic) methods also are available for similar conditions. The one promoted here is the one I raised in passing and then only by example, namely, triangulation. I have come to see triangulation—that is, the use of multiple sources, methods and, in this book, theories—as a very powerful way to analyze and make recommendations about exceptionally complex issues. In fact, I now take the metanarrative in narrative policy analysis to be a form of triangulation (i.e., we converge on the metanarrative by triangulating from the two or more positions that are contrary or contradictory to each other). In policy analysis, the only conversation stopper worse than "That's just a story..." is "...but it's more complex than that"; the aim of *Narrative Policy Analysis*, and now *Taking Complexity Seriously*, is to show how very far analysts can go when addressing both stories and complexity head on. (That third major conversation stopper, "Oh, but that issue is much too big to deal with now...," is to be addressed in the next and last book—*Poverty, Defense and the Environment*—of what I think of as my "Reviving Policy Analysis" trilogy.)

The third driver of *Taking Complexity Seriously* reflects the fact I have run a center on sustainable development and teach classes on that subject and on advanced policy analysis methods for the Berkeley campus of the University of California. When I found myself working on sustainable development, it quickly became clear I needed my own answers to the four big questions—What is sustainable development? Why is it an issue? Ideally, what should be done? and Practically, what can be done? The added attraction of the topic is that "sustainable development" is easily one of the handful of pre-eminently difficult policy issues that face us—you, me, everyone. Sustainable development became an ideal case study to illustrate the practical uses of triangulation for what is of concern to many of us professionally and personally.

None of these reasons, however, account for *Taking Complexity Seriously* as much as my office. "Policy analyst" is the occupation I've been writing on those airport departure and arrival cards since graduating from Michigan in 1972. For these past twenty-five years, I have worked with remarkable people, but never of the caliber and energy of the group assembled at Berkeley's Center for Sustainable Resource Development. David Zilberman, Janet Supár, Peggy Nguyen (along with our Urban Environmentalism class) and, most important, Andrés Davidovits and Ken Hall, have given this book its purpose and made it a pleasure to write. For that reason, it is dedicated to them.

Much of the material in *Taking Complexity Seriously* represents revised journal articles. Earlier versions of Part I and Part III appeared as Emery Roe (1994), "Against Power, For The Politics of Complexity," in <u>Transition</u> 62 and Emery Roe (1994), "The Sharp Edge of the Sword: Reply to My Critics," in a special symposium on my article "Against Power, For the of Complexity," in <u>Transition</u> 64. I am grateful to the <u>Transition</u> managing editor and Duke University Press for permission to use this material. An earlier version of Chapter 1 is reprinted from <u>Ecological Economics</u>, Vol. 16, Emery Roe (1996), "Sustainable Development and Girardian Economics," pp. 87-93, with the kind permission of Elsevier Science-NL, Sara Burgerhartstraat 25, 1055 KV

Amsterdam, The Netherlands. Chapter 2 is based on an article that appeared as Emery Roe (1996), "Sustainable Development and Cultural Theory," in International Journal of Sustainable Development and World Ecology, Vol. 3. I thank Parthenon Publishing for permission to use the article. Chapter 3 appeared originally as Emery Roe (1996), "Critical Theory, Sustainable Development and Populism," in Telos, Vol. 28, No. 1 (Spring). I thank the Telos editor, Paul Piccone, for permission to reprint. An earlier version of Chapter 4 appeared as Emery Roe (1997), "Sustainable Development and the Local Justice Framework," in Philosophy and Social Criticism, Vol. 23. I am grateful to SAGE Publications, Ltd. for permission to reprint. Parts of Chapter 5 will appear as A. Damodaran and Emery Roe (forthcoming), "Theorising to Explain and Triangulating to Explain Away? The Art and Non-Art of Multimethod Policy," in Economic and Political Weekly. I also am grateful to Springer-Verlag New York for permission to use in Part III an earlier article that appeared as Emery Roe (1996), "Why Ecosystem Management Can't Work Without Social Science: An Example from the California Northern Spotted Owl Controversy," in Environmental Management, Vol. 20, No. 5.

Finally, allow me to sum up by setting the stage for Part I with the concluding stanza of "Corsons Inlet" by today's poet of science and complexity, A.R. Ammons:

> I see narrow orders, limited tightness, but will
> not run to that easy victory:
>   still around the looser, wider forces work:
>   I will try
>   to fasten into order enlarging grasps of disorder, widening
> scope, but enjoying the freedom that
> Scope eludes my grasp, that there is no finality of vision,
> that I have perceived nothing completely,
>   that tomorrow a new walk is a new walk.

REFERENCES

Ammons, A.R. (1986), *The Selected Poems: Expanded Edition*, W.W. Norton & Company, New York.

Arnold, Peter and Robert Leone (1992), "Different Stories, Common Themes: The Role of Line Managers in the Public Sector Case Literature," *Journal of Policy Analysis and Management* 11(2), pp. 348-357.

Averch, Harvey and Milan Dluhy (1992), "Teaching Public Administration, Public Management, and Policy Analysis: Convergence or Divergence in the Masters Core," *Journal of Policy Analysis and Management* 11(3), pp. 541-551.

Bardach, Eugene (1987), "From Practitioner Wisdom to Scholarly Knowledge and Back Again," *Journal of Policy Analysis and Management* 7(1), pp. 188-199.

Barzelay, Michael and Linda Kaboolian (1990), "Structural Metaphors and Public Management Education," *Journal of Policy Analysis and Management* 9(4), pp. 599-610.

Behn, Robert (1986), "An Opportunity, Not a Problem," *Journal of Policy Analysis and Management* 5(3), pp. 639-645.

Behn, Robert (1987), "The Nature of Knowledge About Public Management: Lessons for Research and Teaching from our Knowledge About Chess and Warfare," *Journal of Policy Analysis and Management* 7(1), pp. 200-212.

Behn, Robert and Regina Brough (1990), "The Four Keys to a Successful Executive-Education Program: Curriculum, Logistics, Marketing, and Faculty," *Journal of Policy Analysis and   Management* 9(1), pp. 138-143.

Brock, Jonathan (1991), "Learning from Experience: Programs for Executives and Some Implications for Policy Schools," *Journal of Policy Analysis and Management* 10(4), pp. 719-733.

Brock, Jonathan (1993), "New Curriculum and Case Project," *Journal of Policy Analysis and Management* 12(2), pp. 420-424.

Brock, Jonathan (1993a), "Breaking the Barriers: A Practical Guide to the Use of Cases and Discussion Teaching," *Journal of Policy Analysis and Management* 12(4), pp. 819-823.

Brock, Jonathan (1994), "An Electronic Hallway: Desktop Case Access Becomes Available," *Journal of Policy Analysis and Management* 13(4), pp. 778-784.

Diver, Colin (1986), "No Compromises," *Journal of Policy Analysis and Management* 5(3), pp. 645-648.

Doron, Gideon (1986), "A Comment: Telling the Big Stories—Policy Responses to Analytical Complexity," *Journal of Policy Analysis and Management* 5(4), pp. 798-802.

Elmore, Richard (1986), "A Sense of Where You Are," *Journal of Policy Analysis and Management* 5(3), pp. 648-654.

Elmore, Richard (1986), "Graduate Education in Public Management: Working the Seams of Government," *Journal of Policy Analysis and Management* 6(1), pp. 69-83.

Elmore, Richard (1989), "How We Teach Is What We Teach," *Journal of Policy Analysis and Management* 8(1), pp. 173-176.

Heifetz, Ronald, Riley Sinder, Alice Jones, Lynn Hodge, and Keith Rowley (1989), "Teaching and Assessing Leadership Courses at the John F. Kennedy School of Government," *Journal of Policy Analysis and Management* 8(3), pp. 536-562.

Leone, Robert (1989), "Teaching Management Without Cases," *Journal of Policy Analysis and Management* 8(4), pp. 704-711.

Lynn, Jr., Laurence (1987), "Public Management: What Do We Know? What Should We Know? And How Will We Know It?," *Journal of Policy Analysis and Management* 7(1), pp. 178-187.

Lynn, Jr., Laurence and Zhang Jiping (1994), "Reforming the Chinese Economy via the Case Method," *Journal of Policy Analysis and Management* 13(3), pp. 582-589.

O'Hare, Michael (1991), "Formal Models and Government: Teaching to Do," *Journal of Policy Analysis and Management* 10(3), pp. 519-541.

Quigley, John and Suzanne Scotchmer (1989), "What Counts? Analysis Counts," *Journal of Policy Analysis and Management* 8(3), pp. 483-489.

Robyn, Dorothy (1987), "What Makes a Good Case?," *Journal of Policy Analysis and Management* 6(2), pp. 292-295.

PART I

ANALYZING POLICY ISSUES OF HIGH UNCERTAINTY AND COMPLEXITY

ANALYZING POLICY ISSUES OF HIGH UNCERTAINTY AND COMPLEXITY

So why is management consultancy doing so well? The answer can be summed up in two words: complexity and uncertainty. Complexity creates confusion; uncertainty creates fear; and both create a booming demand for outside advice.

The Economist, March 22 1997

LAST GAS STATION BEFORE THE DESERT

One of the more famous typologies in organization theory comes from the work of Thompson and Tuden (Thompson 1967):

		Ends	
		Agreement	*Lack of Agreement*
Means	*Agreement*	Computation	Bargaining
	Lack of Agreement	Majority judgment	Inspiration

Simply put, when there is agreement over means and ends, computation is often the preferred decisionmaking strategy. When there is agreement over the means but not the ends, bargaining is preferred. When goals are agreed upon but not the means to achieve them, then majority judgment becomes the way the parties frequently proceed to decision. And when no agreement over means and ends exist, then the parties have only inspiration to get themselves out of the mess in which they find themselves.

Without commenting on the merits of the other three, it is that "inspiration" which sticks in the throat. *Analysis* was once all the vogue, but increasingly inspiration is there to challenge it, and in a big way. We need more inspiration in organizations, John Brandl advises us (Brandl 1989, 491-493). For their part, Feldman and March (1988, 425) mark the recent shift to an "organizational

3

action that is legitimized by symbols of ambiguity and intuition," in short—inspiration.

Taking Complexity Seriously recommends the opposite. When there is no agreement over means and ends, analysis still remains the viable option. Issues of extreme uncertainty and complexity can be analyzed quite effectively without falling back to inspiration alone. The analytic methods required for these sometimes desperate situations are, however, not those taught in most of our methods courses and seminars. Yet we proceed ahead today as if the old methods will get us across this complex public policy terrain. When those peter out, we seek inspiration, a.k.a. luck, leadership, intuition, or the high octane of political will, to propel us the rest of the journey. What we missed before starting are those fairly recent developments in analytic methods that could fuel us farther, more reliably.

This book is about that last gas station before the desert, and the "we" are the various audiences—both domestic and overseas—of *Taking Complexity Seriously*:

- practitioners, particularly policy analysts and planners, searching for more useful ways to analyze highly complex and uncertain policy issues for which the direct application of conventional analytic methods (e.g., statistics, microeconomics, planning theory) has proven increasingly inadequate or unsatisfactory;

- policymakers, experts, or members of the public at large seeking a new slant on the highly topical issue of how to think through and do sustainable development;

- upper division undergraduates and graduate students in the social and natural sciences, particularly in interdisciplinary enterprises, such as public policy, environmental studies, science and technology programs, development studies, and/or social science methods courses;

- university students interested in (i) recent developments in economics (Girardian economics), anthropology (cultural theory), or political philosophy (Critical theory and local justice framework) and/or (ii) specific applications of the related work of René Girard, Aaron Wildavsky and Mary Douglas, Paul Piccone, and Jon Elster; and

- applied social scientists, natural scientists or policy academics interested in science and environmental controversies (in this case, the debate over sustainable development), new analytic methods for complexity analysis (triangulation), and/or innovative applications of four relatively powerful approaches to understanding key social phenomena.

In order for my audiences to better appreciate the challenge of complexity analysis, another typology from organization theory, which will be used throughout the rest of this book, needs to be introduced. According to the sociologist, Charles Perrow (1984), a tightly-coupled system is (1) highly time-dependent in not allowing for delays or unexpected contingencies; (2) fairly invariant in terms of the sequence of activities required (i.e., B depends upon A having happened first); (3) inflexible in the way its objectives are by and large achieved (not only is the sequence of specific activities restricted, but there is only one way to achieve the overall goal desired); and (4) characterized by little slack and resources available to tolerate delays, stoppages and the unexpected when they do occur. In loosely-coupled systems, delays are not only possible, but common; sequences of activities are by no means invariant (e.g., in a university it does not much matter when some course requirements are met before getting a degree); many ways to achieve a common goal are available; and sufficient slack exists to tolerate a degree of waste without imperiling system survival in the process. Both tightly- and loosely-coupled systems can be, in turn, complexly or linearly interactive. Complexly interactive systems are those with unfamiliar, unplanned or unexpected sequences of activities that often are not visible or immediately comprehensible. The sequences in a linearly interactive system are by contrast much more familiar and expected and are quite visible and comprehensible, even if unplanned or unintended. The dimensions of coupling and interaction produces a typology of four cells:

		Interaction	
		Linear	*Complex*
	Tight	1	2
Coupling	*Loose*	3	4

Many of us find ourselves more and more in Cells 2 and 4 policy situations, thinking they really are or could be the good old days of Cell 1. Indeed, today's analytic challenge is to avoid deluding ourselves into believing that the best way to adapt to complexity is through thinking only in linear, tightly coupled terms. The linearity of regression analysis, the tight coupling of benefit streams in cost-benefit analysis, even the "basic regularities" that are said to drive Complexity or Chaos Theory (e.g., Lewin 1992; Johnson 1997) are supreme examples of Cell 1 thinking. While we will occasionally distinguish between Cell 2 and Cell 4 situations in the rest of the book, for the most part what distinguishes both from the other cells is that we simply don't know the full causality at work in a Cells 2 and 4 world. If in this kind of world, Cell 1 thinking can get us only so far—indeed, exclusively Cell 1 analysis in Cell 2 and 4 circumstances is one reason why policymaking has been made so intractable—then just what kind of analysis can we undertake when focusing solely on extremely complex policy issues?

This book's chief answer is triangulation—that is, use of multiple methods, procedures and/or theories to converge on what should or can be done for the complex issue in question. As Part II is devoted to an extended case study of how to triangulate on a highly complex policy issue (sustainable development), the remainder of Part I focuses on other important ways and approaches to analyzing Cells 2 and 4 policy issues. Although *Taking Complexity Seriously* focuses on the environmental and natural resource arena, the methods discussed are intended to apply to all complex policy topics. Part III concludes the book by drawing out the rather important implications that the triangulation has for sustainable development in the field, particularly for what is called "ecosystem management" there. As we will see, the Cell 1 world of "power" is far too impoverished for a Cell 2 and 4 "politics of complexity."

SO WHAT'S WRONG WITH CELL 1 THINKING?

The good news is that we aren't being duped—years of research, money and analysis have amply demonstrated that many policy issues, and not just in the environmental arena, are truly complex. The bad news is that conventional research and analysis often do not help in rectifying the issues. How can we analyze highly uncertain, complex and divisive policy problems, such as those in environmental management, when their truth-value or tractability cannot be determined through conventional means or under the time and financial constraints we currently operate? If analytic choice is in some fundamental sense to be based on informed analysis, how can we make such choices when the information to be analyzed and its implications are so unclear or meager?[1]

The Answer

The essence of analytic choice is to underwrite and stabilize the assumptions for decisionmaking in the face of high uncertainty and complexity. If we can understand what preconditions must be in place for policymakers to take decisions on highly controversial issues without simplifying them, we can then design and carry out the analysis and research in ways that accommodate these preconditions. What, then, is meant by "underwrite," "stabilize," "the assumptions for decisionmaking," and "in the face of uncertainty and complexity"?

First, though, our prior question still has to be answered.

To Repeat, Just What Is Wrong With Cell 1 Thinking?

In many cases, nothing. There are myriad occasions when policy research and analysis can be undertaken satisfactorily through cost-benefit analysis, random sample surveys, regression analysis and other well-tried techniques. Moreover, bureaucracies are replete with Cell 1 situations. Indeed, the principal reason why deadlines and crisis management are so popular in complex organizations

---

[1] The issue of what makes for intelligent action in a disorderly world is raised in March and Olsen (1986, 23-29).

is that they give to these situations a linearity and tight coupling that they would not otherwise have (March and Olsen 1986).

The difficulty arises when conventional research methods have exhausted themselves only to confirm the policy issues are really uncertain and complex. In these cases, we require techniques that frame uncertainty and complexity in ways that we can do something about them, without at the same time supposing that the uncertainty and complexity can be dismissed, avoided or otherwise dispelled. To introduce this book's mantra: We require *both* conventional and novel analytic methods to *triangulate* from as many different directions as possible on what we could be doing better in the face of an issue whose empirical merits remain unknown, not agreed upon, or both. From my view, the current methods debate between "quantitative versus qualitative," "reductionistic versus holistic," and "positivist versus postpositivist" approaches to analysis have missed the point altogether: In Cells 2 and 4 you need all of the methods, and more, when triangulating on issues of high complexity and uncertainty.

The Ideal Cell 1 Analysis in Policymaking

The ideal model for policy research is frequently assumed to have up to five components typically undertaken in a linear and tightly coupled fashion: first, a baseline survey identifying policy needs (with "policy" defined broadly to include projects, programs, strategies, plans and administrative reforms); second, rational planning to identify options to meet these needs; third, cost-benefit analysis, or something like it, to determine the optimal option; four, implementation of the preferred option as planned; and last, once policy has been implemented, a full-scale evaluation should be undertaken, preferably using control groups and random sampling, to assess the net impact of the policy in meeting the identified needs against the counterfactual of what would have happened had not the policy been introduced. The ideal model has many variants, but they share the basic assumption that such elements make for good research, and good research is necessary for good policy. Examples of the model pushed into practice are many.[2]

Calling the model "ideal" is NOT intended to make for an easy-to-knock strawperson. The model is NOT be criticized because holes can be punched into such terms as "optimal," "rational," or "planning." Nor is the ideal model criticized because it is "reductionistic," "quantitative," and "positivist." NOTHING in this book repudiates the ideal model or any of its elements, particularly as one of the book's central messages is that you require all the research methods you can get your hands on when triangulating.

---

[2] See Social Science Research Agenda Project (1991, 1-25) and SCR 43 Task Force (1989, ES1-24).

The Real Problem

. . . with the ideal model stems directly from its palpable shortcomings in underwriting and stabilizing the assumptions for decisionmaking in the face of high uncertainty and complexity. While Cell 1 thinking works in Cell 1 situations, it has definite limitations when the policy issue is extremely complex and uncertain. Policy researchers themselves have understood and grappled with these limitations.[3] Moreover, the limitations may be many, but their principal consequence is easily stated. When it comes to policy issues, such as you find in many resource and environmental controversies, the ideal model too often finds that issue to be more complex and uncertain than initially thought. Policymakers want answers, but researchers and analysts end up finding policymakers do not even know the right questions.[4]

To be specific: baseline surveys frequently find more needs than can be met; planning ends up fragmented in an government of multiple and conflicting roles, decision processes, agencies, policymakers and issues, each having different analytical requirements; cost-benefit analysis and related techniques are themselves methodologically debatable; implementation rarely takes place as planned and delays are endemic; and evaluations of existing policies often prove difficult to undertake according to experimental designs, or where randomization and control groups are used, findings prove considerably less robust than required for policy purposes.[5] Add these problems together and you have the reason why the policy world is so "messy" for both policy researcher and analyst. This is a world where one simply cannot conclude, "Well, you know, had they done the analysis beforehand, the policy wouldn't have turned out to be the disaster it was." Even though *Taking Complexity Seriously* focuses on ways to improve analysis of complex policy issues, the reader should be under no Cell 1 illusion that the link between analysis and choice or analyst and policymaker is tightly coupled and linear nor that good analysis is a sufficient, let alone necessary, condition for good policy. Instead, the reason why you want improved analyses of complex issues is to ensure you, as the analyst or policymaker, are better prepared for the few policy windows

---

[3] Nathan (1988, 20-43 and 69-96); the *Journal of Human Resources* interchange with Henry Aaron, Edward Gramlich, Eric Hanushek, James Heckman and Aaron Wildavsky on Nathan's book (1990, 276-311); Rossi and Wright (1984, 331-351); Cohen and Weiss (1977, 67-83).

[4] See the Senate Office of Research's criticism (1989, 1-4) of the SCR 43 Task Force report.

[5] See as well Kingdon (1984, 75-94 and 205-218); Wilson (1989, 72-89 and 154-175); Meltsner (1986/1976, 14-49); Pressman and Wildavsky ([1984/1973, 94-110); World Bank (1992, 78-119); Chambers (1983, 51-55); *Journal of Policy Analysis and Management* interchange with John Quigley et al (1989, 482-503); *JPAM* interchange with Donald Rosenthal et al (1992 and 1993, 116-129 and 195-199).

described below where such analyses matter, and matter profoundly. More in a moment.

Note that these problems with the ideal model stem not just from different values and material interests in society. If it were simply a matter of the latter, then homogenizing values or equalizing assets could be a goal. Even if that were possible, it still would not be enough to resolve the resource management and environmental controversies that interest this book. An insistent message here is that the policy world—domestic and overseas—is truly multi-dimensionally complex, uncertain and plural. The sustainable development controversy raises difficult issues, whatever one's values or material interests. That controversy, like so many others, is complex all the way down. In this way, the most telling criticism to be leveled against the Cell 1 model of policy research is not that it is wrong, but that it makes policy analysis seem much easier—in short, more tightly coupled and sequential—than it really is or could be for many controversies.

SO WHAT DO WE MEAN BY "UNDERWRITING AND STABILIZING THE ASSUMPTIONS FOR DECISIONMAKING IN THE FACE OF UNCERTAINTY AND COMPLEXITY"?

To return then to the earlier question: If the essence of analytic choice is to underwrite and stabilize the assumptions for decisionmaking in the face of high uncertainty and complexity, just what do these terms means?

"Underwrite"

Nothing underwrites (i.e., establishes or certifies) the belief that useful decisions can actually be made in the face of extreme uncertainty and complexity except our prior faith and caring enough to believe this is so. There is no rational or axiomatic proof that says one must take a decision and in this way while in a complex and uncertain world. Some would prefer to believe that concepts and theories underwrite our belief in analytic choice, but it is more accurate to say they stabilize the assumptions for decisionmaking. What underwrites your belief in analytic choice is your leap of faith—there is no better term for it—that this is the most desirable way to proceed. There are, of course, other ways to proceed, most notably the earlier inspired choice. Admittedly, this book's premise—analysis not just inspiration, triangulation not just leadership—is only one among many valid positions in a plural world.[6]

---

[6] Leaders, of course, can make intelligent choices as well as inspired ones, though intelligence is not the only attribute—not even the most important one—many look for in leaders. Heifetz et al (1989, 539): "A successful leader has influence because of personal traits, such as bearing, dominance, intelligence, and tone of voice." We also are learning the ways in which inspired and analytic choice are related, as when "hunches" that precede analysis may help render that subsequent analysis more effective (e.g.,

"Stabilize"

Policymakers rely on a battery of means to stabilize (i.e., fix and steady) their ability to make decisions when matters are chaotic or opaque around them. Most prominent are price signals, policies, rules of thumb, bureaucratic war stories, crisis scenarios, and regulations that guide policymakers in making decisions when decisionmaking matters.

But how are these means of stabilization any different than the (too) tightly coupled, linear Cell 1 thinking? Are they not just simplifications that can and often do create more problems in the end? Yes and no. The issue is, if you will, more complex than simplification. To see why, consider the literature's twelve "admonitions" to the analyst:

(1) Cell 1 thinking is unavoidable in a Cells 2 and 4 world. When analyzing any difficult policy issue, one most often must start and end with, *What's the story here?* Telling stories—more formally, scenarios and argumentation having beginnings, middles and ends or premises and conclusions—are the pre-eminent way people stabilize the assumptions for policymaking in highly uncertain and complex situations. When you cannot determine causality, you have to analyze people's argumentation and scenarios over that causality. (Narrative policy analysis [Roe, 1994] provides one such framework for analysis.) Moreover, your policy recommendations will be cast in a narrative format to make them comprehensible not just to your client but to yourself. Indeed, that is why simplicity, clarity and transparency are so valued in policy analysis—they are precisely the values prized in good storytelling.[7]

(2) But "What's the story here?" only begins and ends the analysis. In between, you must triangulate from as many different directions (theories, frameworks, experiences) as possible on the answer to that question, i.e., what's the storyline and argumentation here. There is, of course, no promise that you will intersect on the problem, let alone a way ahead. As Majone (1989, 67) puts it, "policy problems carry no guarantee that there exist correct solutions against which the analyst's conclusions could be checked."

(3) Confidence in research findings is built up in triangulating when you offset the major biases in your analysis, while remaining sensitive to biases in the argumentation you are analyzing. (Bias embraces

---

Bechara et al, 1997).

[7] See Neustadt and May (1986, 91-110) and Bardach (1992, section on "Tell the Story," 27-30). The importance of stories in policy analysis has been mentioned in *JPAM* by Krieger (1988, 410); Sykes (1990, 302); and Averch and Dluhy (1992, 549), among others. Simplicity, clarity and transparency in policy analysis are emphasized by Henry Aaron in the *Journal of Human Resources* interchange (1990, 276-311).

- theories stabilize
  assumptions for
  decision-making

fallacies, threats to validity and pitfalls in analysis.) You cannot eliminate bias in analysis, but reducing or offsetting bias is one of the few things that analysts *are able to do*, and do well, when all else around them is much less certain, simple or resolvable.[8]

(4) One way to get a storyline straight is to reduce the gap between research users and gatherers, that is, between decisionmakers and researchers. Researchers—whether bureaucrats, villagers or others— render their information more policy relevant when they cease to be just its gatherers and become its users. When this happens, research and analysis have a better chance of becoming timely feedback, prompt learning and rapid adaptive responses.[9] Analysis, in this case, *is* the intervention.

(5) A different way to admit complexity into the storyline and argumentation is to think in terms of, be prepared for, and know ways to accommodate **surprise**. The "most important characteristic of complexity," Demchak concludes (1991, 17), is "that complexity produces surprise."[10]

(6) If the objective is to be ready for surprise, you must think counternarratively by looking for the counter-expected event(s). If you are told a –> b –> c, can you think of a plausible scenario, either where (not-a) –> (not-b) –> c or where a –> b –> (not-c)? Once you have a rival hypothesis, ask of it: Is it *surprising*? Is it *desirable*? Can it be treated *seriously*? In addition to better preparing you for surprise, the virtue of counternarrative thinking is that, even if the rival hypothesis is shown not now to be the case, it can still serve as a possible policy option (i.e., counterscenario) for the future to the extent that it is plausible, desirable and serious.[11]

(7) Indeed, one good way to show you are taking complexity seriously is to replace terms, such as rational, efficient, optimal and second-best,

---

[8] Sorensen (1992, 256-266); Chambers (1983, 13-23); Chambers (1992, 14); Campbell and Stanley (1963, 5-13); Majone (1989, 55-68); and, especially, Nathan (1988, 69-96).

[9] See Feldman and March (1988) and Chambers (1992) on the gap between information gatherers and users. The notion that the two should become the same turns on its head Hanushek's research hierarchy (1990) where disciplines feed practitioners, but rarely the other way round.

[10] See also Demchak (1991, Chapter 2); Shackle (1969, 277-285; and 1988, 1-7); and Gross and Rayner (1985, Chapter Two, passim).

[11] Thinking in terms of counter-expected events, adverse scenarios, and counternarratives is discussed in Shackle (1969, 284), Bardach (1992, 16-17); and Roe (1994), respectively.

with those of "plausible," "desirable," "serious," "surprising," and their synonyms.[12]

(8) Taking complexity seriously in an analysis also means, Beware analogies from the past. They are all-too-often the worst kind of Cell 1 thinking in a Cells 2 and 4 world. Why? Because these analogies are tightly coupled and linear, whereas a Cell 2 and 4 world can be defined as those situations where knowing the past helps you less and less in moving up the learning curve.[13] Complexity is always easier to increase than these analogies imply. Values along any one of complexity's three dimensions can change at the drop of a hat, i.e., more components to a system are added, differentiation among the components is increased, and/or the interdependence of components is intensified.[14]

(9) Another way to admit complexity into the analysis is to ask, What is this a case of that I am researching or analyzing?[15] Is this policy a variable to be manipulated or is it a case in its own right? This is why triangulation and offsetting bias in analysis is essential. Without them, you jeopardize understanding just what it is you are investigating is an instance of.

(10) If you take complexity seriously, there will be times when the best policy analysis you can do—repeat, *the very best analysis*—is to evaluate each case on its own merits. In practical terms, this means balancing the five evaluative criteria analysts and policymakers commonly use to assess policy performance: Did performance of the policy match its stated goals? Did performance conform to some ideal, whether stated or not in the policy document? How did performance fare against the counterfactual? Did the policy do better than similar policies implemented elsewhere? and Could performance have been undertaken better (e.g., more efficiently or cost-effectively)? In some

---

[12] Such terms also point to the reason why people think that stories, particularly theories and policies, not only stabilize decisionmaking but underwrite it as well. When an argument is a plausible, desirable, and serious one, the temptation is to say it is, in a word, believable. When that happens, stories become as much the post hoc rationalization for our original leap of faith to make a choice as they are the opportunity to make that choice an intelligent one.

[13] A similar point is made in Demchak (1991, 18). On drawing analogies from the past, see Neustadt and May (1986).

[14] Demchak (1991, 28).

[15] Ragin (1992, 1-17). Ragin, of course, is not the only one to focus on the question, "What is this case of?". Rosenau and Durfee (1995) recommend that "one must be predisposed to ask about every event, every situation, or every observed phenomenon, 'Of what is it an instance?'" We return to the Ragin framework in Part III.

instances, you may be able to treat each criterion equally, and simply aggregate answers to determine how much of a success or failure the policy was. In more complex situations, you will have to take each criterion into account but weigh each differently—and when you do that, you are using your best judgment and treating each case on its own merits.[16]

(11) Taking complexity seriously also means acknowledging that the usefulness of analysis and policy research cannot be predicated solely on its success in explaining the past or predicting the future. Why? Because finding a possible cause of something that has already happened is always easier than predicting the probable consequences of something that has yet to happen. Moreover, the reasons why we cannot predict or plan better for the future are virtually the same reasons why we do not learn more from the past, i.e., in both cases we require stability in objectives, institutional memory, multiple reserves and low environmental uncertainty—all of which are woefully missing in a Cells 2 and 4 world. Yes, policy research and analysis do have an engineering function, but one more imperative and difficult than commonly supposed. *It is to identify what we are able to do, not just what needs to be done.* Indeed, our sense of complexity comes in large part from our increasing inability to determine just what we have been and are now able to do. "My sense is that we are in fact backsliding, for we do not know *how to do* the very things we are doing," according to Sartori (1989, 393): "What we have is [social] engineering without engineers."

(12) Finally, the most subtle way to take complexity seriously is to distinguish beforehand pluralism from relativism. Relativism is "I like coffee, you like tea; it all depends on your point of view." Pluralism is "I like coffee, you like tea; but I understand your reasons for liking tea, though here are my reasons for still preferring coffee." It is because the world is plural that issues can be analyzed as uncertain and complex. If it were all relative, none of that would matter, as you could not escape your own point of view. A pluralist world means uncertainty and complexity are objectively real. They are not discursive formations. But since things are really uncertain and complex, objectivity itself is highly circumscribed. For all intents and purposes, the closest that you, the analyst, get to objectivity in such situations is to identify just whose uncertainty and complexity you are analyzing. Is it the uncertainty of those you are researching, when they express doubts or identify unknowns? Is the uncertainty yours, when confronted by the conflicting certainties of those you are working for or addressing? Or is the uncertainty more systemic and visible only at the issue level? One way in which we answer these questions is to appeal to a metalanguage that allows us to compare and contrast plurally different

---

[16] As Chambers (1992, 50) puts it, "When you have to, use your own best judgment!"

worldviews, where the metalanguage is itself not relative to any one of these worldviews. Triangulation is part of this metalanguage, which insists that often incommensurable perspectives can be objectively compared and contrasted for the purposes of drawing conclusions. In this way, triangulation is not so much a call for methodological pluralism as it is a method for assessing and evaluating pluralism.[17]

What do these admonitions add up to methodologically? As that superb economist, George Shackle (1969, 277) has pointed out, in an uncertain world policies must be periodically revised and redirected, or else they are not policies. Policies stabilize the assumptions for analytic choice, but once choice is taken, surprise is inevitable, thereby necessitating revision and eventual overhaul of the policy in question. Policies must have built into their implementation the capacity to be redesigned through, e.g., effective monitoring and evaluation mechanisms. Thus, no matter how useless conventional (Cell 1) social science research has been for policymakers, *research and analysis turns out to be absolutely essential in policymaking situations of many unknowns and high controversy.* The problem is to find the level and form in which to make it useful, and even then there are no guarantees.

"Assumptions for Decisionmaking"

Evidence from the literature, like that above, is of no use unless the analyst directs it to ensuring his or her analysis is used in actual decisionmaking. You can't criticize your way to policy relevance; you have to work for it.

But what makes for policy relevance? It turns out that research is not policy relevant just because it is research on a major public policy issue. Rather more practically, analysis is policy relevant only when those assumptions of access to decisionmakers who use the research to make decisions that are actually implemented are realized. This is familiar ground, but let me briefly say something first about the assumptions and then about the decisionmaking.

Assumptions The four assumptions for policy relevant research are that you, the analyst, have *access* to a decisionmaker, that she treats your research *seriously*, that she *uses* the research for making a decision, and that the decision is actually *implemented* in light of your research findings.[18] There are, to repeat, no guarantees any of this holds. You have to work at meeting these assumptions and, as with all policy analysts, you not only have to be *persistent*, you must also be able to *compress* analysis into minutes of advice.[19] As Kingdon

---

[17] On the difference between pluralism and relativism, see Berlin (1991, 70-90) and Putnam (1983, 234-240.

[18] See, for example, Heller (1986).

[19] On the need for compression, see Katz (1993); on persistence, see March and Olsen (1986, 26).

(1984, Chapter 9) underscores, your world of analysis will have only a few *policy windows*, i.e., a limited number of opportunities when these four assumptions are all met, in whole or in part. (We will return to the importance of policy windows in Part III.) If policymakers have to fight to get heard and to be taken seriously, where, if they win, they frequently do so only at the last minute, with one vote to spare, and then only after a great deal of effort, why ever should the lot of the analyst be any different?

Decisionmaking   The four assumptions imply that policy research is done for a client, a public or private sector decisionmaker. There are, of course, analysts and researchers who believe that their client is "society," "the public interest," or "the body politic."[20] Here the research and analysis are treated as if they were a message in the bottle, thrown onto the open seas in the hope that someone, somewhere, sometime soon, will retrieve the message and realize its urgency and importance. It is difficult, however, to maintain that fiction of a common public interest or society for very long when your policy world is fundamentally plural, uncertain and rarely simple. A "coping agency" unable to measure its outputs and outcomes cannot be expected to demand, as a client, the same kinds of analysis, let alone research, as does that agency capable of measuring its outputs and outcomes (what James Q. Wilson calls a "production agency"), for example.[21]

"In the Face of Uncertainty and Complexity"

By this point, readers have asked more than once, "Just what does he mean by 'uncertainty' and 'complexity'?" Social scientists typically distinguish uncertainty from risk and ignorance,[22] but for most analysts the three merge into the analyst's lack of knowledge about what matters. What matters, moreover, is often the issue's complexity. To many analysts, complexity is the issue's internal intricacy and/or its interdependence with other policy issues. In fact, as was sketched earlier, system complexity typically has three elements—the number of components in a system, the degree of differentiation between these components, and the degree of interdependence among components. Again, there are more sophisticated definitions of uncertainty and complexity, but for the purposes of this book the two can be defined as above.

As we also saw a moment ago, the interplay of complexity and uncertainty produces *surprise*. Accordingly, if the essence of analytic choice is to underwrite and stabilize the assumptions for decisionmaking in the face of complexity and uncertainty, the essence of such choice is *to make decisions in*

---

[20] Meltsner (1986/1976, 37) and Hallberg (1992, 4).

[21] Wilson (1989, 158-171).

[22] See, for example, Viscusi (1992, 153-154).

the face of surprise.[23] The implications go well beyond the necessity of policy analysts and policymakers to be better prepared for and able to accommodate surprise in their Cells 2 and 4 world. *It implies that if the policy world is full of surprises, so too should analysis itself be surprising.* Surprises attract attention; they can even make their own policy windows. If you are able to triangulate on a (not the) solution to what others have already concluded is intractable or nearly so, then you have thereby produced a result deserving the decisionmaker's second look. We shall see just such an example of "surprising solutions" in Part II's case study on the sustainable development controversy.

SO WHAT DOES THIS ALL MEAN FOR POLICYMAKING?

We are now in a position to redefine what complex public policy must be for issues like those populating environmental and natural resource controversies. Hallberg (1992, 4) defines public policy conventionally as "a course of action consciously chosen from among the available alternatives and designed to fulfill a specified need or set of needs of the body politic; it includes specification of the means by which this course of action is to be implemented."

As we just saw, equally important as a course of action is the ability to revise and redirect action in light of surprise. These actions may be consciously chosen, but it is better to say they are triangulated on. Moreover, the triangulation is from different directions, not just or even primarily from among existing alternatives. Courses of action may be designed to meet needs, but it is better to say they are designed to capitalize on the various abilities and capabilities of real clients or the very real people being analyzed, not something as unremittingly opaque as "the body politic". To that end, policies must specify the means whereby action is to be achieved through the use of policy-relevant analysis and other implementation mechanisms (i.e., research itself must be part of implementation). Bringing these qualifications together, we redefine policy as follows:

> Under conditions of uncertainty and complexity, public policy are courses of action *and* their revision or redirection in light of surprise, both of which are to be triangulated upon from different directions and methodologies; such policies are designed to capitalize on the various abilities and capabilities of the client, decisionmakers or persons concerned and include specification of the means by which this revision and redirection is to be achieved through the use of policy-relevant research and other forms of analysis in implementation.

In a fundamental sense, what sets redefined policy apart from its conventional counterpart is that the former is our response to the core problem of surprise as

---

[23] Demchak (1991) and Shackle (1969 and 1988) are a good introduction to this topic.

it moves through time and across issues. In policy controversies, revision and overhaul are not something that happens to policy; they are what make it policy in the first place.[24]

SO WHERE TO NOW?

When the policy world is increasingly complex and where policy itself has to be redefined as above, there we require the new approaches and ways to proceed touched upon above. The one that this book recommends, triangulation, is demonstrated in Part II's case study, to which we now turn.

REFERENCES

Attfield, C.L.F., D. Demery and N.W. Duck (1991 [1985]), *Rational Expectations in Macroeconomics: An Introduction to Theory and Evidence*, Second Edition, Blackwell, Oxford, United Kingdom.

Averch, Harvey and Milan Dluhy (1992), "Teaching Public Administration, Public Management, and Policy Analysis: Convergence or Divergence in the Masters Core," *Journal of Policy Analysis and Management* 11(3), pp. 541-551.

Bardach, Eugene (1992), "Problem-Solving in the Public Sector," a handout, pp. 1-30 (with appendix), Graduate School of Public Policy, University of California, Berkeley.

---

[24] The argument that effective policy includes the means of its revision addresses the oft-noted "paradox" that the more uncertain conditions are, the less surprising are our responses [see Earl and Kay, 1992]. We try to stabilize our decisionmaking in the face of surprise by reducing the element of surprise in our own decisionmaking routines. Given half a chance, as Karl Mannheim put it, bureaucrats turn policy issues into administrative problems—and, with good reason. The stakes are high in correctly specifying this issue of what makes for effective policy. For example, the case being made here is that effective policy is a course of action which includes the means for revising itself in the face of uncertainty. Rational expectations theory would seem to argue just the opposite: Effective policies are those that create uncertainty. Its well-known policy ineffectiveness proposition contends that only policies unanticipated by economic agents can have real effects. In contrast preannounced policy changes will be ineffective, while effective policies are inadvisable because of the uncertainty they engender (Milberg, 1993, 270). Attfield, Demery and Duck (1991, 97-98) explain policy ineffectiveness this way: "...to be effective these random changes in government expenditure or the money supply, or whatever, would have to be unpredictable; and that means unpredictable to the government too, for if the government could predict them, then so could others. Their effect would be not to stabilize output but to increase the fluctuation of outputs..." The argument made above is that the means of revision have to be there if policy is to be effective, not that these means of revision have to be known or fully specified *ex ante*.

Bechara, Antoine, Hanna Damasio, Daniel Tranel, and Antonio R. Damasio (1997), "Deciding Advantageously Before Knowing the Advantageous Strategy," *Science*, Vol. 275, 28 February, pp. 1293-1295.

Berlin, Isaiah (1991), "Alleged Relativism in Eighteenth-Century European Thought," in his *The Crooked Timber of Humanity* (Henry Hardy, ed), Alfred Knopf, New York, pp. 70-90.

Brandl, John (1989), " Interchange by John Quigley, Suzanne Scotchmer, John Brandl, Robert Behn and Isabel Sawhill on cost-benefit analysis and its alternatives," *Journal of Policy Analysis and Management* 8(3), pp. 482-503.

Campbell, Donald and Julian Stanley (1963) *Experimental and Quasi-Experimental Designs for Research*, Rand McNally College Publishing Co., Chicago.

Chambers, Robert (1983), *Rural Development: Putting the Last First*, Longman, London.

Chambers, Robert (1992), "Rural Appraisal: Rapid, Relaxed and Participatory," IDS Discussion Paper 311, Sussex, pp. 1-67.

Cohen, David and Janet Weiss (1977), "Social Science and Social Policy: Schools and Race," in *Using Social Research in Public Policy Making* (Carol Weiss, ed), Lexington Books, Toronto, pp. 67-83.

Demchak, Chris (1991), *Military Organizations, Complex Machines*, Cornell University Press, Ithaca.

Earl, Peter and Neil Kay (1992), "How Economists Can Accept Shackle's Critique of Economic Doctrines Without Arguing Themselves Out of Their Jobs," in *Joan Robinson (1903-1983) and George Shackle (1903-1992)* (Mark Blaug, ed), An Elgar Reference Collection, Aldershot, England.

Feldman, Martha and James March (1988), "Information in Organizations as Signal and Symbol," in James March's *Decisions and Organizations*, Basil Blackwell, pp. 409-428.

Gross, Jonathan and Steve Rayner (1985), *Measuring Culture*, Columbia University Press, New York, pp. 1-56.

Hallberg, M.C. (1992), *Policy for American Agriculture*, Iowa State University Press, Ames.

Hanushek, Eric (1990), "The Policy Research Markets," *Journal of Policy Analysis and Management* 9(2), pp. 146-154.

Heifetz, Ronald, Riley Sinder, Alice Jones, Lynn Hodge, and Keith Rowley (1989), "Teaching and Assessing Leadership Courses at the John F. Kennedy School of Government," *Journal of Policy Analysis and Management* 8(3), pp. 536-562.

Heller, Frank (1986), "Introduction and Overview," in *The Use and Abuse of Social Science* (Frank Heller, ed), Sage Publications, London, pp. 1-18.

Holliday, Ian (1992), "On Michael Oakeshott," *Government and Opposition* 27(2), 131-147.

Journal of Human Resources (1990) Interchange by Henry Aaron, Edward Gramlich, Eric Hanushek, James Heckman and Aaron Wildavsky, 25(2), pp. 276-311.

Journal of Policy Analysis and Management (1989) Interchange by John Quigley, Suzanne Scotchmer, John Brandl, Robert Behn and Isabel Sawhill on cost-benefit analysis and its alternatives, 8(3), pp. 482-503.

Journal of Policy Analysis and Management (1992 and 1993) Interchange by Donald Rosenthal, Robert Nelson, Raymond Kopp, and John Quiggin on including or excluding existence values in cost-benefit analysis, 11(1), pp. 116-129; and 12(1), pp. 195-199.

Katz, James (1993), "Science, Technology, and Congress," *Society* 30(4), pp. 41-50.

Kingdon, J.W. (1984), *Agendas, Alternatives, and Public Policies* , Little/Brown, Boston.

Krieger, Martin (1988), "The Inner Game of Writing," *Journal of Policy Analysis and Management* 7(2), pp. 408-416:

Lewin, Roger (1992), *Complexity: Life at the Edge of Chaos*, Macmillan Publishing Company, New York.

Majone, Giandomenico (1989) *Evidence, Argument and Persuasion in the Policy Process*, Yale University Press, New Haven, pp. 55-68.

March, James and Johan Olsen (1986), "Garbage Can Models of Decision Making in Organizations," in *Ambiguity and Command* (James March et al, eds), Pitman Publishing, Marshfield, MA, pp. 23-29.

Meltsner, Arnold (1986 [1976]) *Policy Analysts in the Bureaucracy*, California University Press, Berkeley.

Milberg, William (1993), "Natural Order and Posmodernism in Economic Thought," *Social Research* 60(2), pp. 255-277.

Nathan, Richard (1988), *Social Science in Government: Uses and Abuses,* Basic Books, NY.

Neustadt, Richard and Ernest May (1986), *Thinking in Time: The Uses of History for Decision-makers*, Free Press, New York.

Perrow, Charles (1984), *Normal Accidents: Living With High Risk Technologies*, Basic Books, New York.

Pressman, Jeffrey and Aaron Wildavsky (1984 [1973]), *Implementation*, University of California Press, Berkeley.

Putnam, Hilary (1983), "Cultural Relativism and Cultural Imperialism," in his *Realism and Reason: Philosophical Papers, Volume 3*, Cambridge University Press, Cambridge, pp. 234-240.

Ragin, Charles (1992), "Introduction: Cases of 'What is a case?'," in *What is a Case?: Exploring the Foundations of Social Inquiry* (Charles Ragin and Howard Becker, eds), Cambridge University Press, Cambridge, pp. 1-17.

Roe, Emery (1994), *Narrative Policy Analysis*, Duke University Press, Durham, NC.

Rosenau, J. and M. Durfee (1995), *Thinking Theory Thoroughly*, Westview Press, Boulder.

Rossi, Peter and James Wright (1984), "Evaluation Research: An Assessment," from *The Annual Review of Sociology*, Annual Reviews Inc, Palo Alto, CA, pp. 331-351.

Sartori, Giovanni (1989), "Undercomprehension," in *Government and Opposition* 24(4), pp. 391-400.

SCR 43 Task Force (1989), "Executive Summary," to its *The Challenge: Latinos in a Changing California*, University of California, Irvine, pp. ES 1-24.

Senate Office of Research (1989) criticism of SCR 43 Task Force Report, pp. 1-4.

Shackle, G.L.S. (1969), *Decision, Order and Time in Human Affairs*, Cambridge University Press, Cambridge, pp. 277-285.

Shackle, G.L.S. (1988), "The Origination of Choice," in his *Business, Time and Thought*, Macmillan, London, pp. 1-7.

Social Science Agricultural Agenda Project (1991), "Executive Summary," to its *Social Science Agricultural Agendas and Strategies* (G. Johnson et al), Michigan State University Press, East Lansing, MI, pp. 1-25.

Sorensen, Roy (1992), *Thought Experiments*, Oxford University Press, New York.

Sykes, Gary (1990), "Learning to Teach With Cases," *Journal of Policy Analysis and Management* 9(2), pp. 297-302.

Thompson, James D. (1967). Organizations in Action, New York, McGraw-Hill.

Viscusi, W. Kip (1992), *Fatal Tradeoffs*, Oxford University Press, New York and Oxford.

Johnson, George (1997), "Researchers on Complexity Ponder What It's All About," *New York Times*, Science Times, Tuesday May 6, B9, B-13.

Wilson, James Q. (1989) *Bureaucracy: What Government Agencies Do and Why They Do It*, Basic Books, New York.

World Bank (1992), *The World Bank and the Environment*, Washington, DC.

PART II

THE SUSTAINABLE DEVELOPMENT CONTROVERSY

INTRODUCTION

The following five chapters are the core of *Taking Complexity Seriously*.

The four chapters—"Sustainable Development and Girardian Economics," "Sustainable Development and Cultural Theory," "Sustainable Development and Critical Theory," and "Sustainable Development and the Local Justice Framework"—answer the same questions: *What is sustainable development? Why is it an issue? Ideally, what needs to be done? Practically, what can actually be done?* Each chapter bases its analysis on the same debate over sustainable development in the journals, *Science* and *Ecological Applications*. That recent debate and its generalizability is summarized in the next section. Part II's concluding chapter—"Triangulating on Sustainable Development"—uses the four approaches to converge on a set of overall answers to each of the questions. Part II is written as if it were a "whodunit," the solution to which the reader doesn't know until all the parties have been questioned on the way to the denouement in the concluding triangulation.

The four approaches are explained in their respective chapters. The reasons for their choice are detailed in Part II's conclusion. To summarize, each provides a powerful, albeit radically different, tack on the complex management issues core to sustainable development. By using diametrically different approaches to intersect on answers to the four questions, we will have increased our confidence as to what sustainable development is all about, given the admitted complexity of the topic.

The methodology and assumptions behind using triangulation in analyzing issues of high uncertainty and complexity are also laid out in Part II's conclusion. A short summary of the criteria used in selecting the four specific approaches—Girardian economics, cultural theory, Critical theory, and the local justice framework—is provided here. Triangulation of theories requires that: (1) each approach should take complexity and uncertainty seriously, i.e., each represents at some fundamental level a theory of uncertainty; (2) the approaches in question should be orthogonal on the dimension of comparison, which, in our case, means each differs fundamentally on the core sustainable development issue of resource management; and (3) each approach defamiliarizes the problem of sustainable development in fresh ways, i.e., more conventional analytical frameworks, such as microeconomic analysis, are often part of the problem being analyzed, which is certainly the case in the sustainable development debate. The four approaches used in the Part II triangulation meet all three criteria.

These approaches are orthogonal (though not perfectly so) in the sense that for any approach's resource management position "a," the other three approaches pose roughly diametrical positions in the form of "not-a," "both a and not-a," and "neither a or not-a." In terms that will be clear by the end of Chapter 5, Critical theory contends there is a hegemonic technical managerial elite of resource managers opposed to local level control, while the local justice framework insists there are cases where the techno-managerial elite can be found locally as well (a; some not-a). In contrast, cultural theory argues that the techno-managerial elite, whether local or not, is just one of several basic cultures of resource management (both a and all not-a), while Girardians will have none of that, insisting that none of these cultures are in any sense permanent, doomed as they are to disappear in crises of undifferentiation (neither a nor not-a). Thus, if these fundamentally divergent approaches do indeed converge (i.e., triangulate) on some answers to our four questions, we as analysts can be fairly confident that we are on to something that should be followed up, as in "No matter from what direction you look at this issue, you're led to the same [conclusion, starting point, problem definition, or other desideratum]." As we will see, convergence and confidence do not mean we have miraculously found "common ground" between approaches that differ so radically from each other. Rather, what we have done, when triangulation is successful, is redescribed a complex issue in new ways that we feel confident to pursue further.

THE SUSTAINABLE DEVELOPMENT DEBATE

In their now well-known *Science* article, "Uncertainty, Resource Exploitation, and Conservation: Lessons from History," Donald Ludwig, Ray Hilborn and Carl Walters (1993) throw considerable cold water on sustainable resource development, that is, the notion of using resources today in ways that do not foreclose on the future's options to use the same resources. The authors suggest that exploitation of natural resources at sustainable levels is impossible, scientific claims to the contrary are not to be trusted, therefore the promise of sustainable development makes little policy sense. In their view (547), "resources are inevitably overexploited, often to the point of collapse or extinction." Even if it were possible, "sustainable exploitation cannot be achieved without first overexploiting the resource," and then it may well be too late, as "overexploitation is not detectable until it is severe and often irreversible" (548, 547). Their conclusion is made all the more potent because Ludwig, Hilborn and Walters have been leading researchers in sustainable resource management.

"Distrust claims of sustainability," Ludwig *et al* advise (548): "Because past resource exploitation has seldom been successful, any new plan that involves claims of sustainability should be suspect." If better resource policy is wanted, then we must recognize that it cannot be based on such claims (549): "Political leaders at levels ranging from world summits to local communities base their policies upon a misguided view of the dynamics of resource exploitation...[T]he scientific community has helped to perpetuate the illusion of sustainable development through scientific and technological progress. Resource problems are not really environmental problems: They are human problems..." Some of the ways in which they are human problems will become clear in the following chapters.

The Ludwig *et al* article was reported in *The New York Times* and led to a firestorm of responses. The following focuses on the article and the rebuttals from well-known ecologists and experts, including Paul Ehrlich, Harold Mooney and Robert Costanza, in the November 1993 volume of *Ecological Applications*. This interchange remains the most detailed, thoughtful and timely confrontation ecologists have had so far on sustainable development. While some respondents in *Ecological Applications* agree with points in the original *Science* piece, no response agrees completely—save on the centrality of uncertainty in resource management.

Ludwig, his co-authors, and their critics admit they start in the midst of pervasive uncertainty and frequent surprise, the hallmarks of Cell 2 and 4 situations. "Confront uncertainty," Ludwig *et al* (549) recommend: "Effective policies are possible under conditions of uncertainty, but they must take uncertainty into account." When uncertainty is ignored, the "consequence of such decision making is a continual series of surprises and failures" (Hilborn and Ludwig 1993, 552).

So too say their critics. "What sustainability does require...is attention to the management of systems under uncertainty," according to one respondent in *Ecological Applications* (Levin 1993, unnumbered). "There is...an inherent unknowability and unpredictability to sustainable development," concludes another (Holling 1993, 554). In the latter's view, "knowledge of the system we deal with is always incomplete. Surprise is inevitable," thus necessitating "flexibility for adaptation to surprises" (553-554). Sustainable development means "the task of environmental leadership is to recognize uncertainty and to weave it into the adjustments of human life that are necessary if environmental quality is to be protected," in the opinion of a third respondent (Lee 1993, 563). For him (563), "[n]atural dynamics are shrouded in uncertainty". To a fourth, sustainable development "acknowledges the existence of true uncertainty rather than ignoring it, and includes mechanisms to safeguard against its potentially harmful effects" (Costanza 1993, 580). "Because everything in ecosystems constantly changes," according to a fifth, "[s]ustainable development is a moving target" (Salwasser, 1993, 588).

The editor of the *Ecological Applications* volume echoes Ludwig *et al* when counseling that "we must not delude ourselves into believing that the issues [of sustainable development] are entirely scientific, or even primarily scientific" (Levin, unnumbered). Critics also agree that human greed and shortsightedness have very real implications for the feasibility of sustainable development (Ludwig *et al*, 548; Lee, 561; Zedler 1993, 578). Still, their critics dissent substantially from Ludwig *et al*, and understanding these differences is a main charge of the following chapters.

Specifically, the critics believe that the management and exploitation of resources at sustainable levels is not only possible, but that in some cases it is actually happening. "No term receives more attention in discussions of environmental management than 'sustainability'," opens the special issue in *Ecological Applications*. Yet, "Is it an appropriate goal for management?," continues the editor (Levin, unnumbered). "Yes" is the answer, if that journal's responses to Ludwig *et al* are the measure.

According to Holling (554), examples from the Everglades of Florida, the forests of New Brunswick, the estuary of Chesapeake Bay, the Great Lakes, and the Baltic Sea demonstrate "flexibility and innovation" in the "redesign and expansion" of resource policy. "Examples of the new technologies arising from scientific knowledge to satisfy the sustainable use objectives are the 'New Forestry'... and the innovative approaches to 'Biodynamic Agriculture'," according to Mooney and Sala (1993, 565). In Rubenstein's view (1993, 585-586), "there is at least one 'real world' situation, African grasslands, where it should be possible to manage biotic resources in a sustainable fashion," given existing information. "We maintain," contrary to Ludwig, Hilborn and Walters, "that the history of fisheries management provides both positive examples of sustainable resource use and lessons for future improvements" (Rosenberg *et al* 1993, 828).

Ludwig, Hilborn and Walters are themselves reluctant to concede that management is entirely impossible. In their view, if resources cannot be managed, needs must be, and what these needs are is quite clear: namely, the need to reduce rapid population growth and excessive use of resources. "Halt human population increase," Ludwig (1993, 556) admonishes, and "Reduce per capita consumption of resources". Their penchant for management is seen in the heading of the second part of the Ludwig *et al* article, "Some Principles of Effective Management," albeit the principles—particularly "Distrust claims of sustainability" and "Include human motivation"—speak to the difficulty of managing resources and what they see as the necessity of addressing needs directly. Ludwig and his co-authors (549) concede it is theoretically possible to manage resources in the face of uncertainty ("There is a well-developed theory of decision-making under uncertainty..."); unfortunately, political leaders as well as other ecologists have decided to ignore this and instead base their policies upon "the misguided view" of the dynamics of resource exploitation.

Differences between Ludwig *et al* and their critics over the possibility and practice of resource management are important because management is central to the sustainable development debate. Sustainability has been defined by the Ecological Society of America as "management practices that will not degrade the exploited system or any adjacent systems" (Meyer and Helfman 1993, 569). All the responses in *Ecological Applications* tie sustainable development to resource management, and often in their very first paragraphs.[1]

The debate looks, in sum, to be primarily over fact. Is it really the case that Chesapeake Bay, the Everglades, or Africa offer hope for, let alone examples of, sustainable development as just defined? We will see, however, that the debate is not primarily over facts, though facts do matter. In particular, there are very different notions of "management in the face of uncertainty and complexity" that are at issue here—and it is in these differences over management that the answers to the four questions that drive this enterprise will

---

[1] "No term receives more attention in discussions of environmental management than 'sustainability'....," begins the editor's introduction to the *Ecological Applications* debate (Levin, *op. cit.*).

be found. We turn now to the four approaches, all of which take complexity and uncertainty very seriously but each of which takes a fundamentally different tack on the management issues as discussed in the *Ecological Applications* debate. In the process of understanding these differences and the virtue of triangulation, the representativeness and applicability of the *Science/Ecological Applications* debate to wider controversy over sustainable development will become evident.

ONE FINAL REMARK, AND A PERSONAL ONE

Allow me to conclude by explaining why I chose the *Science* and *Ecological Applications* debate as the focus of the case study. In writing *Taking Complexity Seriously*, I wanted an issue where there was little doubt about its complexity and about the need for new ways of thinking through the issue. My criterion of issue choice was that it should be complex all the way down, with few participants to the case study arguing that the solutions are straightforward and simple.

Sustainable development is just such a controversy whose complexity has been accepted by most, if not all, its major parties from the outset. I did not want *Taking Complexity Seriously* to be a book that stands or falls on developing a procedure for separating complex issues from those that are not, since there can be no such once-and-for-all, hard-and-fast (read, Cell 1) procedure for a truly complex world. Fortunately for the purposes of this book but unfortunately for policy analysts and policymakers, there are those policy issues whose complexities are well-recognized, albeit not well-understood, in the sense they are manifestly complex in having many components that are highly differentiated and interdependent. Sustainable development is one, and emphatically so. The vast majority of people see no easy answers here. Furthermore, I have parsed the controversy through the lens of ecologists, because ecology is one discipline that accepts complexity from the very outset as well, at least at the ecosystem level. Ecosystems are paradigmatic examples of habitats whose constituent parts—functions, services, structure and populations—are variously coupled (i.e., a mix of tight and loose coupling) as well as complexly interactive. There are, of course, the few who think sustainable development and ecology are a simple matter of, e.g., no more cars, no more urban sprawl, no more cutting old growth. Others would dismiss sustainable development or ecosystem management as the latest fad and jargon. But, again, the vast majority of people do not see sustainable development and its applications this way. There are hard issues here, whatever the buzzword, but issues that have answers, and surprising ones at that, precisely because they take complexity seriously.

1 SUSTAINABLE DEVELOPMENT AND GIRARDIAN ECONOMICS

What is sustainable development? Why is it an issue? Ideally, what needs to be done? and *Practically, what can be done?* are answered in this chapter by relying on the framework of Girardian economics developed by French economist, André Orléan, and based on René Girard's model of mimetic rivalry under conditions of high uncertainty.

PRELIMINARIES

A good entry point into a Girardian analysis of the sustainable development debate is found in *Ecological Applications*, when Donald Ludwig expands points he and his co-authors, Ray Hilborn and Carl Walters, raised with their original *Science* article.

In "Environmental Sustainability: Magic, Science, and Religion in Natural Resource Management," Ludwig (1993) argues there is a great deal of magic and precious little science in the claims that we can manage natural resources sustainably. To distinguish between what is magic and what is science, Ludwig relies on a passage from the anthropologist Bronislaw Malinowski's *Magic, Science and Religion*:

> *Science, even as represented by the primitive knowledge of savage man, is based upon the normal universal experience of everyday life, experience won in man's struggle with nature for his subsistence and safety, founded upon observation, fixed by reason. Magic is based on specific experience of emotional states in which man observes not nature but himself, in which the truth is revealed not by reason but by the play of emotions upon the human organism. Science is founded upon the conviction that experience, effort, and reason are valid; magic on the belief that hope cannot fail nor desire deceive. The theories of knowledge are dictated by logic, those of magic by the association of ideas under the influence of desire.*

Ludwig (556) continues with his own words: "The practice of resource management fits this definition of magic: it substitutes the word for the deed. As

29

long as human desires are unlimited we shall invent magical theories [like "sustainable development"] in an attempt to reconcile the irreconcilable."

For Ludwig, the real problem with sustainable—that is, magical—development is that many, if not most, of these strategies do not address the fundamental issue of unlimited human desires. You do not need theories of sustainable development to tell you that we must, in Ludwig's terms, "Halt population increase" and "Reduce per capita human consumption of resources" (see also Meyer and Helfman, 570). In his view, the urgent problem not being addressed is the obvious one of too many people, each desiring too many things. Individual wants and needs aggregated over all individuals is what is killing us, if we are to believe Ludwig. Thus, if sustainable development is to move beyond illusion, human desires must be reduced directly, both at the aggregate level and in individual terms.

I want to step back for a moment and return to Malinowski and his term, "desire." How did he originally define it? A few paragraphs before the above excerpt and in a passage Ludwig does not quote, Malinowski (1954, 85) says:

> Now in all his social and worldly ambitions, in all his strivings to catch good fortune and trap propitious luck, man moves in an atmosphere of rivalry, of envy, and of spite. For luck, possessions, even health, are matters of degree and of comparison, and if your neighbor owns more cattle, more wives, more health, and more power than yourself, you feel dwarfed in all you own and all you are. And such is human nature that a man's desire is as much satisfied by the thwarting of others as by the advancement of himself. To this sociological play of desire and counter-desire, of ambition and spite, of success and envy, there corresponds the play of magic and counter-magic, or of magic white and black.

That is, desire is mimetic (imitative and envious) for Malinowski, where magic is how we articulate the engendered rivalry—and the most ambitious theory of such rivalry is that of René Girard. For the purposes of this chapter, the fundamental problem posed by human desire is not that it is aggregate and individual in the Ludwig sense, but that it is mimetic and rivalrous in the Girardian sense.

We now explore how a Girardian understanding of mimetic desire reconfigures the sustainable development controversy. Since a number of the responses in the *Ecological Applications* volume see a conflict between economic growth and sustainable development (Meyer and Helfman and Mangel *et al*), I use the most developed Girardian framework to date for explaining economic behavior, namely, that of André Orléan (1988, 1989, 1992a, and 1992b). In seeking to explain the origins of money, market conventions and speculation, Orléan provides what is essentially a model of economic booms and busts driven by the instability of desire and imitative rivalry. We turn now to his Girardian model, which is discussed first, then placed within related work in economics, and thereafter its implications for the sustainable development debate are drawn.

THE GIRARDIAN MODEL

For Orléan, economic agents start in pervasive uncertainty. Part of the uncertainty is the instability of human desire: Desire for old commodities gives way to desires for new, previously unimagined, goods and services; consumption and production (just as savings and investment) are always transformed in ways that create rather than satisfy desire. Another part of the uncertainty is human ignorance and the sheer unpredictability of events: In situations where the past and present provide little or no guide to the future, market signals (prices, interest rates, and fluctuations in the money supply) prove increasingly unhelpful in making economic decisions. Our response to these uncertainties? Desire becomes mimetic. Each of us decides to imitate the behavior of the other(s).

"How then in this state of radical uncertainty, of the disappearance of every reference point, do the agents orientate themselves? Imitation constitutes this rationality 'in the final instance'" (Orléan 1988, 104). If a market ends up saying you knew more than I, then I'm better off having imitated you; if the market ends up proving us to be equally ignorant, then I am no worse off than before having imitated you. "In a situation of total uncertainty, when I know nothing about what to expect next, copying someone else improves my performance. Either this other person shares my ignorance and my expected outcome is the same, or the other person does turn out to know something, in which case my imitating him improved my situation" (Orléan 1989, 77).

Such mimetic behavior is found throughout the economy. Panic selling, spiraling inflation, overheated art markets, computerized trading, pyramid schemes and speculation (I sell when you sell, buy when you buy) are just some of the many instances of imitative economic behavior. These frenzies occur on the consumption and production as well as the savings and investment sides of the economy. Even "neutral" prices and money take on mimetic properties under these conditions (more below).

Imitative behavior in the face of economic uncertainty poses from the outset several problems. First, the resulting desires and expectations remain crucially incomplete, for mine are a function of yours and are not mine alone. "Like the Girardian individual, the subject in the marketplace never views his desire head-on; he always reads it obliquely in the gaze of others" (Orléan 1988, 102). Second, surprise and the unexpected are never far behind when I become aware of desires I never before had through envying and patterning my behavior on others I do not fully know (Orléan 1988). Accordingly, the uncertainty in which we find ourselves intensifies.

Third, because desire is incomplete and open to the unforeseen, a new desire emerges in each of us: the desire for wealth. Only by possessing wealth can one approximate desires and expectations that are complete and sufficient in themselves. "Wealth is that principle that Girard calls desire's ultimate goal, the possession of which would finally allow the subject to accede to self-sufficiency" (Ibid, 102). My possession of wealth means not having to imitate or depend on others, as all desires are now open to me "on my own terms;" it means that when surprises happen, they no longer need matter as before. Wealth is a way to decouple ourselves from an otherwise turbulent task environment. Consequently, because I am not as wealthy as others (i.e., my desires are imitative in a way theirs need not be), I envy them their wealth, their greater self-sufficiency. "The determination of wealth is thus the fundamental economic problem" within the Girardian framework (Op Cit).

Knowing just what wealth is in situations of high economic uncertainty is no easy matter, however. At one level, the more uncertain things are, the more wealth we desire and the more that desire matters: "The desire for wealth, as we have defined it, does not possess any *a priori* point of resolution. It is indefinite, floating. It is proportional to the uncertainty that gnaws away at the economic agents" (Ibid, 103). At another level, the specification of wealth itself is problematic in such situations: Once wealth "ceases to be identified with the instituted money, [economic agents] no longer know behind which mask it is hiding. Stocks, real estate, gold, foreign currencies, primary commodities, etc. attract the anxious attention of individuals looking for likely refuges from the 'terrible oscillations of chance'" (Op Cit).

What does this all add up to? In Girardian terms, a crisis of undifferentiation. Uncertainty becomes everywhere pervasive and intensified; economic behavior grows more and more imitative and alike; money and other measures of wealth become less and less useful in distinguishing the wealthy from those who are not; and wealth itself becomes increasingly desired as it becomes increasingly difficult to determine. Markets undergoing crises of undifferentiation are instances where we do not know enough to distinguish satisficing from maximizing or the second-best from the Pareto-optimal, and where no one is clearly right or where everyone thinks they are. Orléan (Ibid, 103-104) summarizes:

> The primal scene on which René Girard invites us to reflect is particularly exemplary; the state of social meaning has been wiped clean; one can see only a crowd of individuals defined by their desire for wealth alone, without merchandise, without money, without government, in the middle of a formless mass of natural contingencies...[T]his primal scene...poses in all its magnitude the central economic question: how can the order of the market emerge from an undifferentiated structure of rivalries?....When the monetary convention was strongly established, everyone was distinguished from everyone else in that he was a concrete producer, attached to specific use values. In speculation monetary mediation vacillates, suspicions arise as to its [the currency's] stability;...distinctions fade away. Individuals are then reduced to their minimal definition, possessors of wealth. This process of undifferentiation transmutes the economic society and its structuring hierarchies into a speculative mob...

The "mob" is to be taken seriously in the Girardian framework. Crises of undifferentiation, such as those in panic selling or ballooning speculative bubbles, lead to mob-like behavior in that the imitation converges on ("polarizes around," in Girardian terms) an unanimous choice by the agents concerned, where in turn that choice becomes self-fulfilling and no longer imitative. Mimetic behavior coalesces around unanimity over what the agents take to be the facts of the case, and belief in these "facts" becomes a kind of self-fulfilling prophecy independent of the original imitative behavior (Orléan 1992a).

The problem with this convergence—be it the bandwagon effect of social science or the mimetic contagion discussed by Girardians—is its utter arbitrariness. Remember, these are situations of pervasive uncertainty. In effect, many beliefs over what are the facts of the case are candidates for unanimity. "The content of the belief in question is indeterminate in the sense that a large number of opinions can become the object of unanimity. As soon as unanimity is achieved, the belief is self-validated, for...the veracity of the belief depends essentially on its being shared by others. Let us emphasize the fact that this kind of self-validation is largely independent of the specific content of the belief" (Ibid, 124). In heated financial markets where everyone is selling at time $t+1$ because, quite simply, most everyone was selling at time t, there is no way to validate that selling save by stating it is what everyone was and is doing.

Once panic selling starts, automated trading clicks in, art prices heat up, hoarding accelerates, speculation becomes rampant, resource exploitation races ahead, it is meaningless to say these actions should be guided, let alone governed, by considerations of Pareto-optimality, sustainable development or some other ideal. You are in classic Cell 2 situations of tight coupling and complex interactions. These actions have already taken on a life of their own; they have their own self-fulfilling logic of reinforcing feedback. External standards of what should be done wither before the spectacle of people increasingly imitating others who may be imitating them as they rush to converge on an action they can all agree upon, but which has nothing to do with what is optimal or sustainable, or if it does, that choice itself is not sustainable. In such circumstances, choice gives way to choice, convention to convention, and the only thing that is sustained is the crisis of undifferentiation that gave rise to choice and convention in the first place.

The dynamic of imitation leading to an arbitrary, but self-fulfilling, unanimity applies to prices and money as well. "There is no more widespread belief in economics than the one that makes prices the expression of inviolable objective realities" (Orléan 1988, 107). Yet in Girardian terms, the unanimous reliance by agents on price and currency indicates nothing more than that the highly imitative behavior of agents has become polarized around (converged on) the belief that price, like money, is what these agents in situations of extreme economic uncertainty must use to measure wealth (more properly, the supply and demand of factors related to wealth). "In studying the logic of the dynamic of undifferentiation, we have demonstrated that it leads to a unanimous polarization on a random object. This choice is none other than the engendering of money" (Ibid, 111). In this view, prices and money are first and foremost social conventions, where conventions "are the social objects created by unanimous mimetic polarization" (Orléan 1989, 91).

When price and money become conventionalized measures of wealth, the effects are profoundly counter to reigning economic theory. In the Girardian framework, complaints about prices being too high are the way we scapegoat prices for causing the uncertainty around us, when in fact it was the uncertainty that gave rise to prices in the first place. Similarly, complaining about high prices simply reinforces market behavior, since scapegoating high prices reinforces the notion that low prices are what will save us.[1] More generally, the model's insistence that a change in wealth creates a change in desires explains why, contrary to conventional microeconomics, an individual's indifference curves change in slope and preference with changes in his or her income.

Most important, the neoclassical notions that money is a neutral medium of exchange, that price and demand (supply) are inversely (directly) related, and that changes in price are due to independent changes in demand (or supply) are turned on their head in the Girardian model. "*In a highly uncertain market, price is the indicator of quality...*: as the price increases, the object becomes more desirable and the demand for it grows in consequence" (Orléan 1988, 110). The more the price of gold rises, i.e., the more valuable it becomes, the more it is desired as a store of value. So too for money. When money measures wealth, the desire for wealth becomes the demand for money, and money, like wealth, creates even more desire and expectation. The more money we have, the more desires and expectations open to us, and the more scarcities we experience (Orléan 1989). In this way, markets create scarcities, or at least our sense of scarcity (Xenos 1989). The convergence on money, like price, enables agents to make distinctions, and the reliance on price and money as measures of wealth creates, however temporarily, differentiation where none existed before: "Through the good graces of mimetic convergence, the principle of wealth is able for a time...[to be] identified with money...By this process difference is borne where before was only a formless mass of objects" (Orléan 1988, 111).

But only for a time. The differentiation that arises is merely the result of what Girardians call the internal mediation of accelerating imitative behavior leading to the emergence of social conventions, in this case the reliance on money and prices. External mediation—e.g., a world where Pareto-optimality acts as an external standard for what agents should do, regardless of what those agents may desire and expect—does not exist in situations of high economic uncertainty (see Orléan 1989 and 1992b). Even the *unanimous* convergence on using price and money as an index of wealth cannot escape the arbitrary nature of the convergence. Convergence is at best a respite, as it does not reduce the uncertainty that gave rise to imitation and convergence in the first place. Uncertainty remains, and thus the desire for wealth is ever urgent, whatever the convention agreed upon to measure wealth. Difference emerges from a crisis of undifferentiation, but eventually that difference is seen for what it truly is—an arbitrary convention that temporarily works for people who can deal with uncertainty and its sequelae, envy and rivalry, in no other way. "The same...logic that under some conditions can generate unanimity can easily degenerate into a cumulative and self-validating process that destroys it" (Orléan 1992a, 126). We face, as a result, repeated booms and busts in

---

[1] From a Girardian perspective, when someone complains about the need for money as the reason why he or she cannot "drop out of the system," he or she is not just making the banal observation that survival in this culture depends on marketed goods. Quite the contrary. In making the observation, the person is in the same instant ratifying the holding of money as a way of differentiating agents from each other and thereby transforming "dropping out" into a kind of undifferentiation and homogenization that holding money helps to forestall.

consumption, production, savings and investment within an economy driven by the instability of human desire and mimetic rivalry. Orléan summarizes the cycles of conventions emerging and eroding away in a persisting crisis of undifferentiation as follows:

> [W]hat mimetic convergence has made, it can just as easily unmake....The monetary crisis is to be interpreted as a resurgence of the desire for wealth. It corresponds to that moment when the central monetary institution ceases to be transcendent but is seen for what it is: one particular sign, contingent, the management of a certain form of economic domination. The conflictiveness proper to the market universe resumes its place to the front of the social stage. Like a gangrene it attacks individual expectations; a new period of instability sets in. Speculation is the mode of expression that most exemplifies this crisis...Thus, one witnesses the constitution of an historical cycle, easily discovered, of creation/destruction of the monetary relationship, with destruction engendering forms that are qualitatively new. Because it establishes the irreducibility of the crisis of undifferentiation, our approach is opposed to technocratic thought: it denies that there can exist a theory of social control. Mimetic contagion is what social engineering strategies always end up tripping over...(Orléan 1988, 112)

Before we return to that bit of social engineering called sustainable development, let us pause to situate Orléan's model of Girardian economic behavior within the economics literature generally.

RELATED WORK IN ECONOMICS

How does the Girardian model fit in with mainstream economics? What do other economists say about mimetic rivalry in the economy, be it imitation of the local "keeping up with the Jones's" or in the contagion form described above?

"Imitative behavior, though largely neglected in economic theory, seems common in real life," in the view of one economist (Conlisk 1980, 276). The neglect in economic theory is much less true today. Conlisk (1980) himself has written on the widespread importance of imitation in economic behavior. Topol (1991) explicitly focuses on mimetic contagion in investment behavior. Shiller and his colleagues (e.g., Shiller 1989; Shiller and Pound 1989) have written about contagion models in investment and the strategic role of imitation among investors. Scharfstein and Stein (1990) and Banerjee (1992, 1993) model "herd behavior" among investors. The critical-mass ("tipping") models of Schelling (1978) and Akerlof (1984), as well as the more recent "informational cascades" model of fads and cultural change developed by Bikhchandani, Hirshleifer and Welch (1992), resonate as well with the Orléan's focus on cycles of differentiation and undifferentiation. Even Hirschman's (1970) classic discussion of the importance of loyalty in economic behavior touches upon core themes raised in the Girardian model. The connections between Orléan's model and the work of these other economists are made clearer in the next section's application of the Girardian model to the sustainable development debate.

The model is, however, clearly at variance with the theory of competitive markets in several respects, the most obvious being the latter's assumption of methodological individualism, where economic agents act on their own and independently of each other. In contrast, as Topol (1991, 799) puts it, the essential feature of mimetic contagion is the correlation of agents' behavior, such that it "is not possible to explain the market price dynamics rigorously with only one-agent statistical properties". From a Girardian perspective, there are no such things as "individual" or "personal" uncertainties driving economic behavior; the uncertainties that matter, at least in the market, are everywhere interpersonal and interdependent. The notions that indifference curves are a function of wealth, that supply creates desire (i.e., markets create scarcity), that the amount desired of a good increases as its price rises, and that the long-run is never possible in mimetic behavior are also contrary to neoclassical economics. Furthermore, the Girardian model holds no brief for the notion that imitative behavior in the market becomes less and less likely as putatively more and better information becomes available (again, "private information" is itself mimetically generated).

Also contrary to conventional microeconomics, "increasing returns to scale" occur in the form of a mimetic contagion gathering momentum as an escalating crisis of undifferentiation until it reaches a critical mass tipping it into polarization around an arbitrary set of new conventions. As Banerjee (1992, 814) puts it, "one would expect increasing returns, which rewards doing what a lot of others are doing, to increase the tendency to herd". Finally and once again in opposition to current economics, markets are way of increasing, rather than reducing, transaction costs among agents (more below). It deserves underscoring that these differences arise largely because the *primum mobile* of Girardian economics, unlike neoclassical economics, is uncertainty, and it is this preoccupation with uncertainty that Girardian economics shares with the earlier "subjectivist" economics of, e.g., G.L.S. Shackle (1972), and with more recent work in what is called post-modern economics (see Doherty, Graham and Malek 1992).

That said, there are surprising parallels between Girardian economics and the neoclassical model. Recall the much-maligned strawman of so many economics textbooks, namely, the assumptions underlying the theory of competitive markets and equilibrium:

> *One is that we are discussing a homogeneous good: Each firm produces the identical product. On both the demand and supply side we assume that consumers and firms are price takers: Each economic agent is too small (relative to the total number of agents in the whole market) to affect the market price by its own actions. (This rules out, for example, firms characterized by increasing returns to scale.) Each consumer is a utility maximizer, and each firm is a profit maximizer. We assume perfect knowledge and costless transactions. These assumptions imply that the law of one price will hold: All trades between buyers and sellers will be conducted at the market price. (Otherwise, all buyers would flock to the seller offering the*

lowest price.) We assume that there are no barriers to entry of firms
to the industry or their exit from it. (Friedman 1984, 376-377;
highlights in the text)

What makes the Girardian model of such interest is that it sets out conditions
under which many of the assumptions could be true empirically. A crisis of
undifferentiation means the homogenization of desire and its objects.
Polarization means there are no barriers before the progress of mimetic
contagion. Prices are the social conventions we polarize around as self-fulfilling
prophecies, that is, we become price-takers because our believing prices
measure value leaves us all taking them the measure of value. Individual agents
are indeed "too small" to affect price precisely because we would not have a
price had not the mass of agents mimetically converged on it. Perfect
information, in its turn, can be seen in the fact that each new imitator of
someone else's behavior brings no new information to the market (Banerjee
1992, 1993; Bikhchandani *et al* 1992), i.e., the only market information is that
price which is known to everyone because we have converged around and
polarized on it.[2] Finally, we are utility and profit maximizers because we are
rational, i.e., imitative in the midst of uncertainty, and thus always in search of
more and more wealth as the basic metric of utility.

THE GIRARDIAN MODEL'S REFORMULATION OF THE SUSTAINABLE DEVELOPMENT DEBATE

How applicable is the Girardian model to issues of sustainable development,
that is, to keeping resource options open for the future? The most famous model
of ecological collapse due to self-interested behavior—the Tragedy of the
Commons (Hardin 1977 [1968])—can easily be reread in Girardian terms of
mimetic rivalry. Reinterpreted this way, the Tragedy of the Commons is pre-
eminently a crisis of undifferentiation, where growing numbers of resource users
increasingly mimic each other in competing for resources they are depleting
and thereby driving to ecological as well as economic collapse. In a Tragedy of
the Commons, the less stable and more uncertain the resource becomes, the
more individuals behave imitatively in a way that accelerates the decline of the
resource, where in the end that resource becomes everywhere barren uniformly
for all (that is, undifferentiated).

Nonetheless, it matters profoundly whether such a collapse is driven in
Girardian terms or strictly by those factors, such as "population growth
exceeding carrying capacity," that are often said to precipitate a Tragedy of the
Commons.

---

[2] Moreover, information is costless in the sense that there is no information to be had in
crises of undifferentiation save the prices we had taken as given, e.g., the social
convention called the "law of one price." As Banerjee puts it, after a certain point in his
contagion model of rumors, "the only source of information is the rumour" (Banerjee
1993, 315).

If readers were to fill in the blank in the following statement, "In situations of high uncertainty, economic agents become increasingly _____," many would write "risk averse" or some such phrase. Few, if any, would write "imitative" or "mimetic," on the premise that "imitation...is the last resort of operators in situations of extreme uncertainty" (Orléan 1989, 78).

The effect of choosing "imitative" rather than "risk averse" is profound, however. In Girardian economics, uncertainty is not a background condition that makes avoiding environmental collapse difficult and risk aversion necessary. Uncertainty actually causes the environmental collapse—both micro and macro—while ensuring that risk aversion is not the only response.[3] Uncertainty is not a constraint on but the cause of collapse, and the mechanism for this causal effect is mimicry. Unless remedial action is taken, the overall effect of this mimicry is to produce an economy that is constantly chasing its own signals, where the resulting cyclic economic behavior ultimately leads to the resource depletion and exhaustion decried by so many.

How can we stop a crisis of undifferentiation—in this case, the ecological collapse of an environment driven to destruction everywhere—from taking place and thereby avoid convergence or polarization around arbitrary "solutions" that do not promote keeping options open for the future? Orléan suggests one answer; the related economic literature suggest more. Their answers can be grouped under four admonitions of decreasing efficacy in forestalling a full-blown crisis of undifferentiation:

- If you can't reduce pervasive uncertainty, the next best alternative is to impede the resulting rivalry ("increase the costs to rivalry");

- If you can't reduce the rivalry, the next best alternative is to impede the associated imitative behavior ("increase the costs to imitation");

- If you can't reduce the rivalry or imitation, the next best alternative is to nurture and prolong differentiation ("decrease the costs to differentiation"); and

- Lastly, if you cannot do any of the above, the only alternative is to slow down or wait out the crisis of undifferentiation ("increase the costs to undifferentiation").

Increase the Costs to Rivalry

Much of "sustainable development" is in effect a call for making natural resources so prohibitively "expensive" for those who compete over them that these resources end up being conserved into the future. To keep resource options

---

[3] This, of course, is not to say that all environmental collapses are Girardian in origin. Some declines in resources have quite clear causal factors and occur in situations of high certainty. Our focus, however, is only on those situations of high uncertainty and causal complexity.

open for tomorrow means restricting market rivalry over resources today, which makes competition (a.k.a. rivalry) "cost" more to all involved, other things being equal. In the Girardian framework, markets are mechanisms to *increase* the transaction costs associated with rivalry, not decrease them as conventional economic theory would have it. Markets are what keep us from killing each other for the goods and services we desire, if simply by transforming us into takers of price rather than thieves of the commodity.

Increasing the costs to rivalry poses a dilemma, however: It fuels, rather than slackens, the desire for wealth. From a Girardian perspective, increasing the costs of rivalry may lessen that rivalry, but these higher costs serve as an incentive for increasing the wealth needed to cover (buffer against) the now-higher costs associated with rivalry. One may be successful in regulating competition in a way that actually conserves natural resources, but the resulting increase in the cost of production based on these resources will induce individuals to seek further means to expand their wealth to compensate for these new expenses. Moreover, one of the few ways to increase wealth is by expanding economic growth—an economic growth which is ostensibly the "problem" to be addressed by sustainable development.

Increase the Costs to Imitation

If you can't reduce the rivalry, then increase the costs of imitative behavior resulting from that rivalry. Such costs to imitation are in fact the motivating pivot for many sustainable development proponents in their insistence that future generations can ill-afford to imitate our current consumption, growth and population patterns. The economics literature indicates at least two ways to reduce costly imitative behavior.

The first is to reduce the occasions for imitation. Scharfstein and Stein (1990) recommend that if imitation is the problem in business meetings (e.g., younger managers deferring to older ones), then all managers could have their say by voting simultaneously or in writing. Alternatively, younger managers could vote or voice their views before others do. In much the same manner, sustainable development advocates bring intergenerational equity as a priority into decisionmaking in order to give future generations the "first vote" in current resource allocation.

The second way to increase the costs of imitation is by expanding the number of "models" available to imitate, thereby making the choice of what to imitate more difficult. Here the premise is that the more things to imitate, the more difficult it is to know what to imitate. The more information there is and the more at variance it is, the harder it is to figure out what to do—other things being equal. For example, if market imitation is a problem, then ensure information is made available to the investor so that she or he "gets to know whether anyone in the population had been aware of the investment opportunity and had chosen not to invest," that is, has chosen not to imitate, when others have (Banerjee 1993, 315). In the same way, the more alternatives there are in practice to current patterns in consumption, production, savings or investment, the more difficult it will be for newcomers to choose to imitate current patterns.

Unfortunately, increasing the costs to imitation in these ways poses very much the same dilemma as increasing the costs of rivalry: From a Girardian perspective, it will make people even more rivalrous. The more intergenerational equity is brought into decisionmaking, the more uncertainty is increased—Just what do future generations want anyway?—and the more wealth people now want in order to buffer themselves against this increased uncertainty. Alternatively, the more alternatives there are to status-quo consumption, the more people will choose to consume something (as in, the more selections before you, the greater the chances of you choosing something).

According to the Girardian model, the fewer choices people have, the more choice people want. The more choices people have, the more people will insist on choosing. Either way, increasing the costs of choice makes people desire ever more choice.

Decrease the Costs to Differentiation

Now relax the assumptions in the Orléan model. You start in the midst of uncertainty, but circumstances are already differentiated, and certainly so (e.g., Shiller and Pound 1989). Imagine pockets of behavior tending toward undifferentiation, but where the pockets are themselves different from each other, i.e., people behave alike within a locality, but localities differ in terms of the behavior exhibited. Orléan and his co-author, Robert Boyer (1992b), show that when this happens, not only can different social conventions emerge, but the conventions evolve and innovate as they diffuse through localities (for a similar finding, see Bikhchandani et al 1992).

The crisis of undifferentiation remains the driver in each locality, but the conventions converged on are drawn or adapted from many different localities. Thus, the more inter-local differentiation, the greater the chances that one of the set of conventions coalesced around will address sustainable development, and thus the greater the chance that sustainable development will be practiced somewhere and all the time, though not everywhere nor always in the same place. Where this is so, each case of sustainable development would have to be treated on its own merits.

How then to create and maintain pockets of local differentiation? While there are no easy answers, what deserves mentioning are two points many would prefer to ignore. First, to what extent are rising population levels, economic growth and per capita consumption—the *bête noire* of sustainable development proponents—creating the very differentiation that is leading to convergence, albeit localized, on sustainable development strategies? The creation of wealth depletes resources, but in the Girardian framework, wealth creates desire and expectation, including for sustainable development. Such is the upshot of the complaint that rich people have the luxury of keeping options alive for the future; the poor have enough trouble keeping alive. If economic growth is The Problem, we must wonder to what extent it has created that very level of socioeconomic differentiation without which people would not care or "need" to keep things sustainable.

The other *bête noire* of sustainable development proponents must also be reassessed when the goal is to maintain an innovating local differentiation: The sovereign nation state may well be one of the most efficient ways to create both differentiation and loyalties to that differentiation. As Hirschman (1970, 80-81) points out, loyalty to one's country grows important as countries become less and less well-differentiated and come to resemble each other more and more. Since the challenge is to encourage pockets of differentiation (otherwise there would be nothing to diffuse), then we should not speak of sovereignty, but of sovereignties; not the nation state, but nation states. Yet many proponents of sustainable development see the multitude of nation states, each claiming its own sovereignty rights, as the primary reason why development is so unsustainable today.

Increase the Costs to Undifferentiation

Assume that the efforts of trying to forestall the crisis of undifferentiation through increasing the costs of rivalry and imitation, and decreasing the costs of differentiation, have failed. (Indeed, reasons have been given why efforts of sustainable development advocates are likely not only to fail, but to actually accelerate the crisis of undifferentiation.) When all else has not succeeded, the final recourse is to wait out the collapse in the hope that the longer you can hold out before undifferentiation becomes total, the more likely that undifferentiation will not be total.

An analogy is useful here. There is a sense in which the older the rumor, the more certain that the rumor mill isn't going to clear up the rumor soon. The later you come to a rumor, the more certain that those who think themselves better informed have already contributed to the rumor mill (Banerjee 1993). In either case, the longer a rumor persists, the more certain you can be it will never go beyond being a rumor. More generally, there may be times when the more imitation seems to go nowhere, the more it will go nowhere; and the longer undifferentiation persists, the longer it will persist without resolution in polarization. Why avoid polarization? Because there is no guarantee that what will be polarized around will be any better than the crisis of undifferentiation being left. To repeat, the object of polarization is always arbitrary in the Girardian framework—that is, the exact antithesis of a sustainable development that is nothing if not long-run planning.

CONCLUSION

The best way to summarize the four admonitions is in the negative: Avoid anything that decreases the costs of rivalry, imitation and undifferentiation; avoid, as well, anything that increases the costs of differentiation. Avoid, in other words, anything that promotes rivalry, anything that makes it easier to imitate one and only one thing wholeheartedly, and anything that makes it harder to be different from each other.

Yet, if we were to follow the recommendations of sustainable development advocates, we would end up increasing rivalry and consumption desires, while working against maintaining local differentiation and assuming nothing worse could happen than the current "crises" of population increase and economic growth. Sadly, then, from within Girardian framework and as has often been said before, sustainable development is an oxymoron—not because development is always unsustainable, but because sustainability cannot be "developed" the way advocates commonly suppose.

We are now in a position to answer the four questions that started the chapter. From a Girardian perspective,

> *What is sustainable development?* It is at best a social convention which for a time underwrites and stabilizes decisionmaking under high uncertainty in a way that its subscribers believe keeps options open for the future.

> *Why is it an issue?* Ultimately it is an issue not because people are overexploiting resources, but because some of them have reached a level of wealth where they now have the desire to stop such exploitation and achieve what they believe to be sustainable development.

> *Ideally, what needs to be done?* Ideally, the uncertainty that drives the crises of undifferentiation people are undergoing should be reduced.

> *But practically what can actually be done?* Buffer or decouple resource systems and their management from the more turbulent environments in which they are embedded. Encourage the evolution and diffusion of more than one kind of sustainable development. Differentiate sustainable development by treating it on a case-by-case basis. Resist globalizing behavior. Last, but never least, nurture both inter-local differentiation and the role of economic growth and sovereignty rights in that differentiation.

2 SUSTAINABLE DEVELOPMENT AND CULTURAL THEORY

What is sustainable development? Why is it an issue? Ideally, what needs to be done? and *Practically, what can be done?* are answered here by relying on the cultural theory of Mary Douglas and her colleagues, especially Aaron Wildavsky and Michael Thompson, both of whom have used their approach to address sustainable development.

THE CULTURAL THEORY MODEL

The basic framework for cultural theory is best summarized in Thompson, Ellis and Wildavsky (1990, 5-7):[1]

> *Our theory has a specific point of departure: the grid-group typology proposed by Mary Douglas. She argues that the variability of an individual's involvement in social life can be adequately captured by two dimensions of sociality: group and grid. Group refers to the extent to which an individual is incorporated into bounded units. The greater the incorporation, the more individual choice is subject to group determination. Grid denotes the degree to which an individual's life is circumscribed by externally imposed prescriptions. The more binding and extensive the scope of prescriptions, the less of life that is open to individual negotiation....[The grid/group typology leads to four basic ways of life.]*
>
> *Strong group boundaries coupled with minimal prescriptions produce social relations that are egalitarian. Because such groups lack (as a consequence of their low grid position) internal role differentiation, relations between group members are ambiguous....When an individual's social environment is*

---

[1] For a specific example of grid/group analysis drawn from the overseas rural development literature, see Roe (1988, 1994a). An excellent introduction to grid/group analysis, with an extended fictional case study explicating the approach, is provided in Gross and Rayner (1985).

43

> *characterized by strong group boundaries and binding*
> *prescriptions, the resulting social relations are <u>hierarchical</u>.*
> *Individuals in this social context are subject to both the control of*
> *other members in the group and the demands of socially imposed*
> *roles....Individuals who are bound by neither group incorporation*
> *nor prescribed roles inhabit an <u>individualistic</u> social context. In*
> *such an environment all boundaries are provisional and subject to*
> *negotiation....People who find themselves subject to binding*
> *prescriptions and are excluded from group membership exemplify*
> *the <u>fatalistic</u> way of life. Fatalists are controlled from without.*

These four cultures are not only competing ways of life, they are the essential pluralism of interdependent cultures: "Each way of life needs each of its rivals, either to make up for its deficiencies, or to exploit, or to define itself against...Were egalitarians to eliminate hierarchists and individualists, for instance, their lack of a target to be against would remove the justification for their strong group boundary and thus undermine their way of life" (Thompson *et al*, 4; see also 96).

Before describing these four[2] basic ways of life, it is important to be clear why multiple ways of sociality are possible. The answer, quite simply, is: uncertainty. "[T]here is always enough irreducible uncertainty in the world for us to be able to bias our convictions this way or that," according to Thompson *et al* (10). "All that cultural theory requires is that there always be some uncertainty of this kind."

For the purposes of *Taking Complexity Seriously*, the four basic cultures can best be described in terms of their different *views about nature*, *strategies for managing needs and resources*, and what each culture takes as "*surprise.*"

Four Views of Nature. From the cultural theory perspective, there must be different views about what nature is because there are different forms of social life. "Ideas of nature, whether physical or human," Thompson *et al* (25) begin, "are socially constructed. What is natural and unnatural is given to individuals by their way of life." Ecologists, according to Thompson *et al* (26), understand this particularly well:

> *In the course of studying managed ecosystems—like forests,*
> *fisheries and grasslands—ecologists discovered that the*
> *interventions of the managing institutions in the ecosystems were*
> *wildly heterogeneous. That is, different managing institutions,*
> *faced with exactly the same sort of situation, did very different*
> *things (some, for instance, started spraying the forest with*
> *insecticide, others stopped)... Though they did different things,*
> *they did not do just anything, and ecologists discovered they could*
> *account for the diversity of institutional response by introducing*
> *into their analysis a number of <u>myths of nature</u>...*

---

[2] A fifth way of life is discussed later in the chapter.

For cultural theory, four "myths of nature" are to be derived from the work of ecologists (Thompson *et al*, 28)—the individualist's Nature Benign, the egalitarian's Nature Ephemeral, the hierarchist's Nature Perverse/Tolerant, and the fatalist's Nature Capricious:

> *Nature Benign gives us global equilibrium [in the view of the individualist]. The world, it tells us, is wonderfully forgiving: No matter what knocks we deliver, the ball will always return to the bottom of the basin. The managing institution can therefore have a laissez-faire attitude. Nature Ephemeral is almost the exact opposite. The world, it tells us, is a terrifyingly unforgiving place and the least jolt may trigger its complete collapse. The managing institution must treat the ecosystem with great care. Nature Perverse/Tolerant is forgiving of most events but is vulnerable to an occasional knocking of the ball over the rim. The managing institution must, therefore, regulate against unusual occurrences. Nature Capricious is a random world. Institutions with this view of nature do not really manage or learn: They just cope with erratic events.*

These distinctions are vital:

> *Nature Benign encourages and justifies trial and error. As long as we do our exuberant, individualistic things, a "hidden hand"...will lead us toward the best possible outcome. But such behavior becomes irresponsibly destructive if nature is ephemeral. Nature Ephemeral requires us to set up effective sanctions to prevent this sort of thing from happening, to join together in celebration of incuriosity and trepidity....*
>
> *Whereas Nature Benign encourages bold experimentation in the face of uncertainty and Nature Ephemeral encourages timorous forbearance, Nature Perverse/Tolerant requires us to ensure that exuberant behavior never goes too far....With Nature Benign, Nature Ephemeral, and Nature Perverse/Tolerant, learning is possible, though each disposes its holders to learn different things (and, thereby, to construct different knowledges). But in the flatland of Nature Capricious, there are no gradients to teach us the difference between...better and worse. Life is, and remains, a lottery. It is luck, not learning, that from time to time brings resources our way.* (Thompson *et al*, 26-28)

Four Management Strategies. Just as one's views of nature are a function of one's cultural bias, so too are what one takes as needs and resources and their management. "Needs and resources," Thompson *et al* (39) maintain, "are socially constructed. The conceptions of needs and resources are, in effect, supplied to the members of a way of life, thereby enabling them to justify that way of life." Each of the four basic cultures has associated with it a "need-and-resource-managing strategy" (Thompson *et al*, 48). Fatalists believe you can manage neither your needs nor your resources; egalitarians believe you can manage your needs, but not your resources; hierarchists believe you can manage your resources, but not your needs; and individualists believe you can manage both your needs and resources. These management strategies, in turn, incorporate their respective views of nature.

Fatalist "Management" Strategy. "The individual who has no scope to manage his needs or his resources really cannot be said to have a management strategy," according to Thompson *et al* (43). "His concern is to cope as best he can with an environment over which he has no control....This response is justified by a view of nature as essentially a lottery...There are clearly plenty of resources out there, but the horn of plenty disgorges in your direction only when it is your lucky day. A matching response to environment can be achieved only by adopting an attitude of fatalism. Putting first things first, fatalists construct a view of nature that operates without rhyme or reason in order to sustain and dignify their way of life."

Egalitarian Management Strategy. "Because you perceive resources to be fixed and you believe people can do nothing about them, your only available strategy is to decrease your needs...But it is no use doing this on your own," in the view of Thompson *et al* (44). They continue: "If resources are fixed and finite, then one person's gain is inevitably another's loss; to be effective, therefore, this need-reducing strategy will have to be followed by everyone... Here the catch phrases refer to 'fragile ecosystems,' 'unsustainable levels of consumption,' 'overloaded arks,' and 'global villages.' By defining resources and raw materials as one and the same, those who follow this strategy are able to insist that all nonrenewable resources are inevitably being depleted and that even renewable ones must be drawn upon frugally...[W]e claim that egalitarians construct the world in such a way as to define resources as fixed. They then choose a needs-and-resources strategy (decrease needs) that justifies their cultural bias and sustains their way of life."

Hierarchist Management Strategy. For the hierarchist, since you cannot do anything about your needs, then the only available strategy is to increase your resources (Thompson *et al*, 44). "Hierarchists, therefore, are not prepared to countenance the possibility that resources are depleting," argues Thompson (1993, 22-23): "Nor conversely, can they swing all the way to the opposite extreme, in which an endless abundance of resources can be secured simply by unleashing the free-for-all of individualistic trial-and-error. The key to resource mobilization, the hierarchist has to insist, lies in the correct matching of the specialised functions of the various divisions of mankind to the fruitful possibilities that exist (by no means everywhere) within a correspondingly stratified and ordered nature....If this is how the world is then environmental management is both necessary and possible. Everything hinges on discovering just where the limits to this pocket of stability lie and then ensuring that people stay on the right side of them."

Individualist Management Strategy. The individualist "has the scope to manage both his needs and resources, and he chooses...to manage his needs and his resources upward to the very limit of his entrepreneurial skills....He is right in the middle of the turbulent stream of competitive individualism, where success comes to those who boldly and skillfully accept the risks—and opportunities— that present themselves there" (Thompson *et al*, 47). For a skillful individualist, this is, in Thompson's words (28), "a world in which, unlike both the egalitarian's and the hierarchist's, individualistic experimentation can harm only the individual, never the totality. Secondly it will have to be a world in which, unlike the fatalist's, it is possible to learn about it by acting vigorously within it...The social construction that sustains the individualist, therefore, can be visualised as a ball in a very deep basin: *Nature Benign*. No matter how we knock the ball about, there is no possibility of it ending up in some other system-state..."

These four different views of nature and their associated needs and management strategies can be summarized in the grid/group typology of the four basic cultures:

Figure 1. Cultures, Views of Nature, and Management Strategies

	Low	High
High (Grid)	FATALIST *Nature Capricious* Can manage neither needs nor resources	HIERARCHIST *Nature Perverse/Tolerant* Can manage resources, but not needs
Low	INDIVIDUALIST *Nature Benign* Can manage both needs and resources	EGALITARIAN *Nature Ephemeral* Can manage needs, but not resources

Group

(based on Thompson, 1993)

Cultural Theory's Typology of Surprises. If one's view of nature and what constitute needs and resources to be managed are socially constructed along the lines of the four cultures, then it should come as no surprise that surprise itself is a function of the goodness of fit between culture and world. "In each way of life just one model of nature is capable of providing the necessary stabilizing forces; all the others would result in that way of life's transformation. Just because one must act in the world in the conviction that it is one particular way, however, it does not follow that the world is that way" (Thompson *et al*,

70). It is when the natural and social constructions of the world are disjunct that surprise comes to the foreground.

For cultural theorists, no event is surprising in itself (Thompson *et al*, 70). Rather, the event is surprising only in relation to a particular set of convictions about how the world is, e.g., convictions about nature or the needs and resources for management consistent with this view of nature. More to the point, the event is surprising only if it is noticed by the holder of that particular set of convictions. What gets noticed is, in turn, a function of the gap between one's view of nature as stipulated by one's culture and the way nature actually is. From these assumptions, Thompson *et al* construct a typology of surprises.

Let the rows in Figure 2 be the actual natures possible and the columns the stipulated natures possible. The cells are filled in with what Thompson *et al* (71) consider to be the major surprise that "would hit the social being, who, stipulating that the world was one way, acted in a world that was, in fact some other way."

Figure 2. A Tyology of Surprises

Stipulated World \ Actual World	I Capricious	II Ephemeral	III Benign	IV Perverse/ Tolerant
I Capricious (Fatalist's myth)		Expected windfalls don't happen	Unexpected runs of good luck	Unexpected runs of good and bad luck
II Ephemeral (Egalitarian's myth)	Caution does not work		Others prosper	Others prosper
III Benign (Individualist's myth)	Skill is not rewarded	Total collapse		Partial collapse
IV Perverse/ Tolerant (Hierarchist's myth)	Unpredictability	Total collapse	Competition	

(Thompson et al 1990, 71)

Since the typology is pivotal to the next section's argument, Thompson *et al*'s description of how surprise leads to changes from one culture to another is quoted at length (71-72, 76; underlines are mine):

> *For instance, if we have assumed a world in which learning is not possible (Nature Capricious), when in fact we occupy a world which is, then we will be slow to pick up the recurrent regularities that are being thrown up around us. And when we do begin to pick up those regularities—when we begin to learn—inevitably, we will find ourselves being eased away from the fatalist's idea of nature and being brought under the thrall of one of the others. Conversely, if we have stipulated a world—Nature Perverse/Tolerant—in which there is a clear boundary between equilibrium and disequilibrium, when the world we occupy is flat and featureless, then try as we may (and we will), we will not be able to obtain the crucial information we need if we are to act rationally...Eventually if we learn from these surprises that learning is not possible, we will find ourselves abandoned by the hierarchist's myth of nature as tolerant (if manipulated by the right people) and perverse (if manipulated by the wrong people) and embraced by the fatalist's vision of nature as capricious....*
>
> *[The shift] from hierarchist to egalitarian...is the path taken by the schismatic: the loyalist who becomes the heretic. In more secular terms, it is the whistleblower.... From hierarchist to fatalist is the most extreme form of a "fall from grace." It corresponds to "defrocking," being "drummed out of the regiment," "discharged with ignominy," "disbarred," "struck off."*

Since empirically we are constantly being surprised—even after having made shifts from one culture to another—the actual world cannot be just one way only: "If the world, everywhere and always, were just one of these [actual] ways, then the surprises that befell all those who insisted that it was some other way would eventually tip them out of their delusions and into that one true world. That this does not happen—that people go on and on being surprised—tells us that the world is never just one way; it is constantly changing" (Thompson *et al*, 72). And with this constant change comes uncertainty, which reinforces and takes us back to where we started with the four basic cultures being a fundamental response to uncertainty.

We are now in a position to reinterpret the sustainable development debate in *Ecological Applications* from a cultural theory perspective.

CULTURAL THEORY'S REFORMULATION OF THE SUSTAINABLE DEVELOPMENT DEBATE

Qualifications follow, but in a nutshell: Ludwig *et al*'s article represents their reluctant shift from a hierarchist to egalitarian position on sustainable development, with strong overtones of a fatalist view of a Nature Capricious. Their critics, in turn, reassert the validity of the hierarchist position that resources and nature can be managed within limits (i.e., sustainably). Ludwig, Hilborn and Walters paint themselves as whistle-blowers on what they see as an increasingly bankrupt hierarchist position,[3] while the critics treat the views of Ludwig and his co-authors as personal opinions outside the pale of what most professional ecologists would conclude. Ludwig *et al* believe that because resources cannot be managed sustainably, needs must be managed directly; their critics believe that, while needs ideally should be managed, resources themselves certainly can be managed sustainably.

It is not difficult to see why the Ludwig *et al* article is called "pessimistic," "negative," and "doomsday" by the critics (Pitelka and Pitelka 1993, 566; Socolow 1993, 581; Slobodkin 1993, 572). There is a kind of fatalism and Nature Capricious at work in Ludwig *et al*'s picture of uncontrolled and uncontrollable exploitation, where "resources are inevitably exploited, often to the point extinction," and "[i]t is more appropriate to think of resources as managing people than the converse" (547). In this way, Hilborn and Ludwig (551) conclude the required learning curve is simply not possible: "What we knew in 1975 is of little use today. We may design a program to learn about the sustainable yield, but by the time the data are collected and we think we know how the system responds to exploitation, it will have changed and our data are obsolete." Nor is it surprising that Ludwig and his co-authors do not believe in sustainable development, as fatalists do not believe in any kind of "managed" development, sustainable or otherwise (Thompson, 18).

By no means does their article concede a full-blooded fatalism. In the authors' view, if resources cannot be managed, needs must be. What these needs are, moreover, is quite clear: "the problems of population growth and excessive use of resources" (Ludwig *et al*, 549). "Halt human population increase," Ludwig (1993, 556) admonishes, and "Reduce per capita consumption of resources":

> *In all likelihood the human population will have to be reduced. How much reduction is ultimately required depends upon our success [in reducing per capita human consumption of resources]...In the case of nonrenewable resources such as petroleum, natural gas, and some metals, the net consumption will*

---

[3] The Ludwig *et al* article starts (1993, 547): "There are currently many plans for sustainable use or sustainable development that are founded upon scientific information and consensus. Such ideas reflect ignorance of the history of resource exploitation".

eventually have to be reduced to zero. In the case of renewable resources, the consumption must be reduced to levels far below the maximum yield level in order to compensate for expected and unexpected variations in abundance. Most importantly, we must allow a margin of safety, since our knowledge of the underlying biological processes is in only a rudimentary form and will remain rudimentary for the foreseeable future.

The passage's focus on needs, non-renewable resources, and caution form the pivot of what is essentially the egalitarian position described earlier. Like all good egalitarians, Ludwig focuses on managing needs directly, on nonrenewable resources being depleted and renewable ones to be drawn upon frugally, and on the necessity of treating such resources and systems with extreme care. For Ludwig *et al* (548), it is "shortsightedness and greed of humans [that] underlie difficulties in managing resources," such that we must "[i]nclude human motivation and responses as part of the system to be studied and managed." They call for "a much more cautious approach to resource exploitation" (Ludwig *et al*, 548). The egalitarian's notion of Nature Ephemeral, where the slightest push can lead to collapse, is glimpsed in their argument that "sustainable development cannot be achieved without first overexploiting the resource" (Ludwig *et al*, 548). Not unexpectedly, some of their critics see the Ludwig *et al* position in egalitarian terms. For Holling (554), "there seems to be something inherently wrong with [their] conclusion...that the only solution is a radical return of humanity to being 'children of nature.'"

Ludwig *et al* are, however, unwilling to give up all hope that resources can be managed. They have not completely recanted their hierarchist position, which leads one critic to identify the article's "two schizoid parts": "The first part, and longest, articulates a deeply held and pessimistic personal belief supported by selected anecdotes...that says human greed and shortsightedness almost always lead to overexploitation and often to collapse of the resource....In contrast, the second part of their article draws on the results of their extensive research and practical experience in analyzing and participating in the science, designs, and management of natural resources....They turn from the conclusion that management is impossible to a set of tested prescriptions for management that are sustainable" (Holling, 553). As already noted, the second part of the Ludwig *et al* article is headed "Some Principles of Effective Management," albeit its principles speak to the difficulty of managing resources and the necessity of addressing needs directly.

The reason why Ludwig, Hilborn and Walters are reluctant to concede completely the hierarchist position that resources can be managed is easy enough to identify. That position is the polestar of so many ecologists and natural scientists everywhere. As Michael Thompson puts it (24):

In the hierarchist's social construction, development is certainly possible, but not everywhere. Development that strays outside the pocket of stability—from tolerance to perversity— will be unsustainable and will have to be identified ahead of time and guarded against. Indicators of sustainability, safe limits, thresholds, critical loads, carrying capacities, statutory

> *assessments of the risks and benefits of new technologies, and a*
> *host of similar concepts and procedures are the means by which*
> *this vital sorting is implemented.*
>   *This, I need hardly point out, is the meaning of sustainable*
> *development that has the widest currency.*

Just such a sorting and assignment position is taken by Ludwig *et al*'s many critics. "Sustainability implies explicit stabilization at levels of resource exploitation that are limited" (Lee, 562); "some highly variable situations can be effectively managed by monitoring combined with regulations" (Slobodkin, 572); and "we should concentrate on defining the borders of a sustainability space by scientifically discarding what can be shown to be nonsustainable" (Fuentes 1993, 576). Such quotes fill the pages of the critics' responses in *Ecological Applications*.

Just as Ludwig and his co-authors are reluctant to give up on the possibility of resource management, so too do their critics recognize that needs themselves should be addressed. "Funding ecological research is not an alternative to taking action to curb human population growth and consumption," argue Meyer and Helfman (570). For Mangel *et al* (574) "sustainable growth of population and resource consumption...is impossible." Pitelka and Pitelka (568) quote approvingly, "Any cause is a lost cause without a reduction in population."

These agreements, however, are not core to the critics' response. Rather, what is core to their opposition to the Ludwig *et al* article is their fundamentally hierarchist orientation that resources—not needs—can be managed. "'Distrust claims of sustainability'...is neither obvious nor especially helpful. While there clearly are problems in determining what levels of exploitation of a resource truly are sustainable, the concept is not only valid, but essential in the long run for the human race" (Pitelka and Pitelka, 568). "The concept of sustainability as a solution to problems of [resource] exploitation is embraced not because of its proven record, but because it offers a viable alternative to uncontrolled exploitation" (Meyer and Helfman, 569). "To maintain biological diversity and options for a sustainable future, societies need to shift the burden of proof from demonstrating that ongoing or planned activities will damage or destroy the resource and have adverse socioeconomic consequences, to demonstrating that ongoing and proposed use will not reduce management options 15-20 yr hence" (Mangel *et al*, 575).

Cultural theory tells us that any one way of life defines itself not only by what it is for, but what it is against, in this case, the other ways of life. We would then expect the hierarchist position not only to reject the egalitarian-cum-fatalist argument of Ludwig, Hilborn and Walters, but also the individualist's view of nature and management strategy. This indeed is what critics do. "At a minimum," Ludwig *et al*'s position, according to Holling (553), "is a refreshing antidote to the cornucopian beliefs of the Julian Simons of the world...who see an infinite capacity for human learning, a limitless ability for people to adapt and substitute". Ehrlich and Daily (1993, 559) conclude, also contrary to the individualist, that "it is difficult to see how 'trial and error' can lead to optimum management of many types of resources. Trial and error can be very dangerous."[4] Egalitarians as well reject the individualist position. Ludwig and his co-authors are emphatic in condemning a laissez-faire, private profit-seeking position of so-called "shortsightedness and greed," i.e., "Wealth or the prospect of wealth generates political and social power that is used to promote unlimited exploitation of resources" (Ludwig et al, 547).

A further test of the explanatory power of the cultural theory model is to return to the typology of surprises. According to Figure 2, the hierarchist, who assumed nature to be tolerant/perverse only to find it actually capricious or ephemeral, would be greatly surprised by nature's unpredictability or total collapse. Having always operated under the assumption that nature could be managed, hierarchists would be shocked to find it uncontrollable or collapsing. Cultural theory, in short, predicts that the Ludwig et al article—if it really represents a shift from an hierarchist to egalitarian-cum-fatalist position—would focus a great deal of its attention on unpredictability and collapse. This indeed the article does, and with a vengeance.

"Confront uncertainty," next to "Distrust claims of sustainability," is the most insistent advice put forth by Ludwig and his co-authors (549). In addition to the "[l]arge levels of natural variability" in ecosystems and "irregular or fluctuating resources," they speak of "our inability to predict," the "great difficulty...in prediction of future events," and the "difficulties that have been experienced in understanding and prediction in fisheries are compounded for even larger scales involved in understanding and predicting phenomena of major concern [to resource management]" (Ludwig et al, passim).

Collapse, too, runs through the article. To repeat, "resources are inevitably overexploited, often to the point of collapse or extinction" (547). "Forests throughout the world have been destroyed," while "fisheries scientists have been unable to control the technique, distribution, and the amount of fishing effort," thereby leading to "the elimination of some substocks" (547). "After the collapse of the Pacific sardine, the Peruvian anchoveta...[had] the most spectacular collapse in the history of fisheries exploitation" (548). "Thus 3000 years of experience and a good scientific understanding of the phenomena, their

---

[4] Other critics do call for "learning through trial and error" (e.g., Socolow, 583), but what they typically mean is the incremental learning through a process of experimentation and re-experimentation called adaptive management (Lee, 563; Meyer and Helfman, 570; see also Ludwig et al, 547). Such incrementalism is consistent with the hierachist position: "In hierarchies," according to Douglas and Wildavsky (1982, 93), "[m]oves are incremental, working with knowledge of levels achieved in the past to do a little more or less. Decision making is remedial (dealing with difficulty) and serial...Problems are solved in sequence." This position of gradual trial-and-error controlled within a hierarchical setting is nicely captured in the Forest Service's definition of adaptive management (USDA, 1993) as the "process of implementing policy decisions as ·cientifically driven management experiments that test predictions and assumptions in management plans, and using the resulting information to improve the plans". Adaptive management was developed in large part by Ludwig's co-authors, Carl Walters and Ray Hilborn, which explains in part why many of their critics treat the recanting Ludwig et al article with such concern and seriousness.

causes, and the appropriate prophylactic measures [such as exist in the case of irrigation] are not sufficient to prevent the misuse and consequent destruction of these resources" (548).

In short, the cultural theory model is amazingly accurate in highlighting the cultural positions and counter-positions, views of nature, management strategies, and notions of surprise that drive the sustainable development debate in the pages of *Ecological Applications* and beyond. What, then, are the theory's implications for recasting sustainable development directly?

CONSEQUENCES FOR A REFORMULATED SUSTAINABLE DEVELOPMENT

The implications are more subtle and considerably less obvious than the reader might first conclude.

Not unexpectedly for a pluralist approach, cultural theory implies that a sustainable development consonant with multiple cultures has a greater likelihood of success than a development which appeals to only one culture. As Thompson *et al* (96-97) put it: "If this reasoning [i.e., cultural theory] is valid, it implies that those political systems that promote a diversity of ways of life are likely to do better than those that repress the requisite variety. Governments need not let a thousand flowers bloom, but they may do well not to nip any of the basic cultural biases in the bud." By this reckoning, the currently dominant conception of sustainable development discussed in the preceding section— hierarchist and, as such, representing primarily only one culture—stands little chance of succeeding in an otherwise plural world. Even some in the *Ecological Applications* debate acknowledge such a conclusion, as when Lee (563) writes: "Pluralistic societies, in which diverse values about economics and resource utilization can be aired, are able to argue their way toward sustainability more effectively than closed societies".

Such reasoning also underlies Thompson's call (27) for "institutional clumsiness" in sustainable development, that is, "of not letting just one (or even an alliance of just two) [cultural] meanings of sustainable development define the problem and its solution." For instance:

> If you are designing policies on the assumption that the individualist's strategy for managing [both] needs and resources is the root-cause of unsustainable development (and many policies are being designed on that assumption) then you will not be giving any attention to ways of encouraging these market outcomes away from environmentally undesirable consequences...

In this view (also Thompson, 16), sustainable development—be it defined solely or largely in hierarchist, individualist or egalitarian terms—will always be misleading. Just as there is no single culture, there is no single meaning of sustainable development. You cannot homogenize development, sustainable or otherwise, in the presence of what are multiple, distinctly heterogeneous cultures and actors. Pluralism must remain the criterion of efficacy.

But how do you do sustainable development "plurally," when the cultures in question are competing and mutually exclusive? The "really big policy question [is] how to encourage the constructive interaction of these plural and ineradicable actors" (Thompson, 17). Cultural theory offers no recipes here, only points of departure.

Cultural theory actually identifies a fifth way of life (and associated view of nature and management strategy), which has yet to be described and which offers one answer to the preceding question (Thompson *et al*, 7, 30-32, 46-47):

> *For a few individuals there is a fifth possible way of life, one in which the individual withdraws from coercive or manipulative social involvement altogether. This is the way of life of the hermit, who escapes social control by refusing to control others or to be controlled by others....*
>
> *But in withdrawing from this fourfold system [of cultures], in which each of the engaged ways of life is endlessly chewing bits off the others, the hermit is sustained not by the unalloyed truth but by his own distinctive myth: Nature Resilient....Because the myths that guide the four engaged ways of life are mutually contradictory, they each define as rational behavior what the others define as irrational. This means that the myth of Nature Resilient can be reached only if these other myths are treated, not as bases for action (because, taken together, they could only furnish a basis for inaction), but as objects of contemplation. Hermits, therefore, do not transcend the other ways of life; they bias themselves away from them....*
>
> *[In the hermit's management strategy] both needs and resources are perceived to be manageable, and the [hermit] has chosen to manage them in such a way that he also has the scope to manage the overlap between them. This means that, unlike the [individualist], his needs will nestle comfortably inside his resources....[The strategy here is to become] autonomous by choosing not to coerce while trying to avoid being coerced by others. For those whose needs are slight (and whose time horizons are short), nature will always provide [i.e., be resilient].*

Notwithstanding terms like "hermit" and "withdrawal," nothing in the fifth way of life necessitates one having to remove oneself from others, as in fleeing to a solitary wood or turning into a recluse. What matters is configuring one's needs and resources so that they coincide in self-sufficiency and autonomy. That this is difficult to do does not mean the fifth way of life ceases to be a model for living sustainably. Indeed, the reader hardly needs a theory of culture to understand why much more meditation on and distancing from cultures-as-usual might well be at the heart of "sustainable development."

That said, What do we do if the fifth way of life is not open to all or even many of us as a model for sustainable development? What if we are left with only the other four cultures from which to choose? To reiterate, an alliance between two cultures is not necessarily the better choice. One need only look to Ludwig *et al*'s egalitarian-cum-fatalist position to see that rather than being the basis for sustainable development, it insists sustainable development is scarcely possible at all.

Are there possible alliances between hierarchists, egalitarians and individualists for the purpose of sustainable development, notwithstanding their status as conflicting ways of life? At least two "alliances" are possible. Examine Figure 1 closely and you will find that quadrants along one diagonal share something in common that those along the other do not—principally, a very different attitude toward trial and error.[5] We have already seen how individualists value trial and error, while fatalists do not. Egalitarians, in contrast, prefer that there be no trials unless one could guarantee against errors, particularly ones perceived to be dangerous, whereas hierarchists are more comfortable undertaking trial and error, albeit in relatively small, incremental steps.[6] In other words, fatalists and egalitarians believe that trial and error is impossible or at minimum potentially very harmful, while individualists and hierarchists believe trial and error is possible, though at times in small steps. Indeed, the *Ecological Applications* articles can be seen as staking out these two diagonals in some detail, suggesting two very different kinds of sustainable development at work depending on what position one takes on the role of trial and error in development generally. Very little attention so far in the sustainable development debate has been given to these different conceptions of sustainable development.

That said, it is in the discussion of alliances between cultures that the subtler implications of cultural theory come to the fore. To appreciate the subtleties, consider the thinking of one of the most original and untiring students of cultural theory, the late Aaron Wildavsky.

Early on, in *Risk and Culture*, Douglas and Wildavsky (1982, 99-100, 180-181) argue that an alliance between hierarchists and individualists has a better chance at stability than egalitarians on their own or in alliance with either hierarchists or individualists, at least when it comes to dealing with related issues of risk and the environment. Their book ends with a discussion on the importance of resilience in responding and adapting to unpredictable change, implying—albeit not explicitly—that the alliance of hierarchist and individualist offers the better way of creating the needed resilience in institutions and organizations.

Wildavsky's later *Searching for Safety* (1988) starts where *Risk and Culture* ends, namely, on the virtue of resilience. Although *Searching for Safety* is not cast in explicit cultural theory categories, it is clear that the valued resilience is now more associated with the entrepreneurial trial-and-error and risk-taking of market competition (read, individualists) than with the bureaucratic

---

[5] In cultural theory analysis, looking at the diagonals can prove very insightful (see Gross and Rayner, 1985; Roe, 1994a).

[6] See also footnote 4.

incrementalism of "trials-with-only-small-errors" (read, hierarchists) or, for that matter, the "no-trials-without-prior-guarantees-against-errors" of many, though not just only, environmentalists (read, egalitarians). As we just saw, by the time of *Cultural Theory* (Thompson, Ellis and Wildavsky 1990), resilience is extended to the fifth way of life, that of the hermit. That said, in one of his last works (and, to my knowledge, the only one where he addressed sustainable development directly), Wildavsky returns to the pluralism criterion of efficacy, arguing we would in the end "be better served by...a pluralist conception of political cultures than uncritically adopting the stance of one way of life to attack another" (1993, 47; such a view can also be found in Thompson *et al*, 255-58).

By pointing to such oscillations in thinking, I do not mean to criticize Wildavsky. Just the opposite. In such views as his are, I believe, the bare outlines of another approach—"model" is too formal—to issues of sustainable development, an approach considerably more complicated than that of any one culture.

Wildavsky was fond of saying, "Just because it's biased doesn't mean it isn't true!" That is, just because each culture has its own cultural bias does not mean that its view of nature is therefore wrong. In cultural theory, stipulated and actual worlds can and do overlap, as we saw earlier, but when they do not overlap, surprise is ever-present. Remember, in cultural theory surprises are the primary way we change our views of the world, and thus shift our culture. We have then to ask, Are surprises distributed equally across cultures? Do some cultures experience a wider gap between stipulated and actual worlds, thereby making them more susceptible to surprise?

Wildavsky saw egalitarians especially prone to surprise in the areas of risk and the environment. His last books related to these topics, *Searching for Safety* (1988), *Safer Power* (with Elizabeth Nichols, forthcoming) and, especially, *But Is It True?* (Wildavsky 1995) are studded with examples of egalitarians, in his case environmentalists, believing one way when the preponderance of evidence suggests otherwise. In Wildavsky's view, egalitarians are continually holding positions—about, *inter alia*, acid rain, global warming, nuclear power, and health and safety regulations—that are surprising not only to others, but should be to egalitarians themselves as well, since these positions prove to have considerably less factual merit than egalitarians suppose.

To repeat, just because all preferences are ultimately cultural and therefore culturally biased does not mean one should or can disregard evidence. Wildavsky summarizes this position under the heading, "Preferences, Yes; Disregard of Evidence, No" in *But Is It True?* (Ibid, 10):

> *Of course, we are not required to be naive [about preferences]. We may conclude that industry-sponsored research is likely to underestimate and environmental groups to overestimate dangers and thus seek to counter these biases...Yet either of these putative extremes may be right. Assuming that "the truth is somewhere in between" is not a logical one and could well be false...What we need is not objectivity in the sense of having*

no views but a willingness to consider different hypotheses and, when considering each, to be guided by evidence.

In this passage as with others, Wildavsky can be read as trying to balance his own cultural preferences for individualism with both a politics that is more pluralistic and a search for truth that applies to all.

Being biased in culture, pluralist in politics, and truth-seeking in the facts, Wildavsky conveys the conviction that best way to defend pluralism is to defend one's own culture, and the best way to defend one's culture (and thereby pluralism) is to insist that the evidence matters, always, whatever the culture. In doing so, one is always vulnerable to surprise and open to shifting from one culture (and associated view of nature and management strategy) to another.

So too can sustainable development be rethought. The permanence of bias in culture, the necessity of political pluralism, the overarching mandate to seek truth, and the openness to surprise and cultural change—one would hope that sustainable development, if it means anything, would mean at least that.

CONCLUSION

We are now in a position to answer the four questions that started this chapter. From a cultural theory perspective,

> *What is sustainable development?*
> It depends on the culture. Because their views of nature and their strategies to manage their needs and resources differ so profoundly, sustainable development is necessarily defined differently by each culture.

> *Why is it an issue?*
> Ultimately it is an issue not because people are overexploiting resources, but because people belong to cultures that are not only different, but conflicting and competing as well. What one culture takes as the answer to uncertainty—its view of nature and its corresponding management strategy—another culture takes as the source of its problems. Wherever there are multiple cultures, sustainable development will be an issue.

> *Ideally, what needs to be done?*
> Ideally, the uncertainty to which culture is a response should be reduced.

But practically what can actually be done?

If you cannot think like a hermit, be biased in favor of your culture (and thus your own definition of sustainable development), recognizing that others are doing the same for theirs. Focus your definition of sustainable development in terms of where you stand on the usefulness of trial and error learning in development generally. Insist on the necessity of pluralism in politics; sustainable development won't be real otherwise. Be truth-seeking, whatever your culture or politics, and expect others to be the same: Sustainable development stands or falls on what is actually happening. Finally, be open to surprise and change: No sustainable development can be or, for that matter, should be once and for all or in one way only.

3 SUSTAINABLE DEVELOPMENT AND CRITICAL THEORY

What is sustainable development? Why is it an issue? Ideally, what needs to be done? and *Practically, what can be done?* are answered through the Critical theory of Paul Piccone and his colleagues at the journal, *Telos*, particularly Tim Luke who has made an extended critique of sustainable development.[1]

THE CRITICAL THEORY MODEL

The armature of Critical theory is the observation that government seeks to rationalize itself and the dominant ideology by means of fostering opposition.[2] Always functionally and sometimes intentionally, bureaucracies engender counter-bureaucracies, each having a vested interest in keeping their "opponents" around as a way of perpetuating themselves. So-called opponents, be they government "versus" the environmentalists, the Left "versus" the Right, or the welfare state "versus" those who would run it better, are part of the same techno-managerial elite that believes it has the solutions to managing our way out of the increasing complexity in which we find ourselves. They differ only in the management solutions they wish effected. Unfortunately but functionally unsurprisingly, complexity increases as these "solutions" give rise to "counter-solutions," with bureaucracies giving rise to counter-bureaucracies and thereby intensifying the belief that a techno-managerial elite is all the more "needed" in order to resolve the increasing complexity.

In *Telos*-speak, the techno-managerial elite (a.k.a. New Class) fosters artificial negativity, which in the beginning gives the appearance of opposition, but in the process supports the elite, and at the end leads to more and more bureaucratic involution. The answer? A negativity that arises organically

---

[1] Critical theory means many things, some of which stand in sharp contrast to the neo-Adornian brand articulated in *Telos* framework. This chapter has kept the journal's capitalization of the term. Although having the same roots, the *Telos* version of Critical theory bears little resemblance to the much better known Habermassian variant.

[2] The following three paragraphs are based on Piccone (1978, 48); Piccone (1987/88, 6-8); and Piccone (1991, 34-35).

outside and in opposition to the increased bureaucratic involution, as in the case of populist movements driving a more loosely-coupled federalism than currently is found in the US. Organic negativity reflects an underlying cultural heterogeneity and particularism—and "cultural heterogeneity is here to stay," as Piccone[3] puts it. More to the point, this heterogeneity and particularity resist all manner of New Class attempts to homogenize human beings into conformity, universalize their problems, coopt and centralize their solutions, and thereby make all individuals abstractly equal when manifestly they are not. Yet, since organic negativity cannot be managed by any techno-managerial elite (it is outside management by resisting it), there can be no blueprints to implement a populism or federalism that better reflects the irreducible cultural heterogeneity and particularity of being. In this way, all of us are left to balance, if not resolve, the increasing complexity of artificial negativity with the open-ended uncertainty of organic negativity.

This simplified gloss of Critical theory has been separated from its historical roots. It is Piccone's view that the rise and persistence of artificial negativity, as a historical phenomenon, has had much to do with a Cold War whose major "opponents" had a great deal in common and mutually to gain in keeping their confrontation going. What ended the Cold War was the collapse of one of the "opponents" as a result of, not US superiority, but the terminal bureaucratization of the Soviet Union. The US is now functionally undergoing just the same involution. Not unexpectedly in this view, the end of the Cold War saw a rise in organic negativity, e.g., secessionist and populist movements, as people realized the ersatz polarization that had kept afloat the bureaucratic behemoths of the Cold War era. That said, much of the above gloss resonates with other similar arguments, including Marcuse's repressive tolerance, Sinfield's more recent entrapment of ideology, and the much-discussed micromanagement cycles of overregulation leading to underregulation leading to even more overregulation and bureaucratic involution.[4]

---

[3] Piccone (1991, 17).

[4] Marcuse (1965) and Sinfield (1992). As for cycles of overregulation and under-regulation, they occur this way: A government (agency, court, legislature) proceeds to adopt more and more regulations to manage what are seen as severe problems of concern. As the number of regulations increase, the regulatory process becomes more time-consuming and the pace of implementing regulations slows down. The slower the implementation and the faster the growth of problems, the more problems appear unregulated. The more underregulation, the greater the pressure on government to micromanage problems by adopting even tougher regulations. But more numerous and stringent regulations slow even further the pace of implementation, and the cycle begins anew. For a sampling of the literature on micromanagement, see Mendeloff (1988, 1-17); Gregory (1989); and Wilson (1989, 241-244 and 366-368).

CRITICAL THEORY'S REFORMULATION OF THE SUSTAINABLE DEVELOPMENT DEBATE

In his *Ecocritique: Contesting the Politics of Nature, Economy and Culture*,[5] Tim Luke neatly captures the New Class aspects of sustainable development. One need only turn to Lester Brown's Worldwatch Institute, a Beltway think-tank preaching sustainable development, to confront the New Class at work. For Luke, "the Worldwatch Institute essentially rearticulates the instrumental rationality of resource managerialism on a global scale in a transnationalized register".[6] The Institute's resource managerialism is nothing if not an aspiring New Class domination of the globe as a whole:

> *Natural resources in the total solar economy of food stocks, fisheries, forest preserves, and grass lands are rhetorically ripped from Nature by the Worldwatch Institute only to be returned by instituting a worldwatch as "environmental resources,," enveloped in accounting procedures and encircled by managerial programs. Worldwatching presumes to know how all of this actually works, and in knowing it, to have mastered all its economic/ecological implications.*[7]

Worldwatch environmentalists and transnational corporations, notwithstanding differences, mirror each other in their shared belief that resources have now to be managed globally. According to Luke, "these 'new class' worldwatchers are, to put a new twist on a popular green slogan, very cynically 'thinking globally' and opportunistically 'acting locally' in order to shore up their own political and economic authority by defending the environment, containing waste, and watching the world in ways that suit their use of power so that their clients and consumers, now on a global scale, will remain passive, dependent and powerless".[8]

Sustainable development advocates differ with others in the New Class only in that they—the sustainable development proponents—see themselves, and not the others, as having answers to how to manage and assemble resources more effectively. In this view, power to manage resources must shift within the New

---

[5] Luke (forthcoming, University of Minnesota Press). For an early discussion of these issues, see the interchange between Barry Commoner and members of the *Telos* group in the section on "The Totally Administered Society," *Telos* 78 (Spring, 1978), pp. 169-184.

[6] Luke (1995, 207).

[7] *Ibid* (213).

[8] *Ibid* (437).

Class, away from those who have managed the earth locally and irresponsibly to those who know how to manage it globally and responsibly:

> *The categories of sustainable development reconstitute Nature—through their recognition of the encirclement of space and matter by a national as well as global economies—as a system of systems that can be dismantled, redesigned, and assembled anew to produce "resources" efficiently and in adequate amounts when and were needed in the modern marketplace without seeing degradation in carrying capacity....Sustainable development discourse, then, might be re-read as a new power/knowledge formation, aiming at accumulating power for comparatively powerless subnational and supranational agencies [like the UN or Worldwatch Institute] through the mobilization of new knowledges about the performance of essentially national economies and states that have exerted their authority to foster development at any cost.[9]*

As such, sustainable development is a New Class form of artificial negativity, where "opponents"—those for a sustainable earth versus those in the earth-degrading transnational corporations—are really in sync with each other in believing that resources can and must be managed on a worldwide scale. Accordingly, much is left unchallenged in this "opposition": "The basic logic of commodification and exchange that animates cases or ecological destruction in the core economies is neglected....[For worldwatchers], 'the world' becomes the site of ecological disasters, economic crises, and political challenges rather than the actual bioregions or countries that are the location of these struggles. Worldwatching trips into the fallacy of misplaced concreteness, globalizing what should be national, regional and local struggles over ecology into general planetary problems".[10]

As a counterweight and corrective to these global aspirations of New Class hegemony, Luke recommends an amalgam of ecological populism, communalistic localism and confederal municipalism, in line with the *Telos* model.[11] It is the organic negativity of "really existing communities," not "sustainable development," that is the greatest threat to New Class resource managerialism, and it is for that reason alone that a globalizing New Class cannot abide the local: "By interposing specialized technical expertise and complex hierarchical organization, which are centered in massive state bureaucracies and transnational networks of corporate production, between the production of most goods and services and their consumption, the new class disempowers people in their families, neighborhoods, and cities..."[12]

---

[9] *Ibid* (271-272).

[10] *Ibid* (228-229). See also the discussion on "analytic tip" in Roe (1994b).

[11] Luke (Chapter 10).

[12] *Ibid* (451).

The very same issues are found in the *Science* and *Ecological Applications* debate. Only here it is the ecologists who are making New Class claims to managing resources better than others. Their claim pulses through the *Ecological Applications* responses. Rubenstein proffers that "effective management can only take place when the best available scientific information is used to inform decisionmaking," adding that "if ecologists really want their scientific knowledge to be taken seriously and to be used in policy making, then ecologists will have to stay involved throughout the decision-making process, from the monitoring of experimental systems to the revising of models".[13] "To be adaptive and contribute to economically and politically effective strategies, science must be integral to the whole management process," in Salwasser's view.[14] The editor of *Ecological Applications* concludes: "What is needed are mechanisms for performing the science that will guide society in making its decisions, and for building bridges between science and decision making".[15] Even when Ludwig *et al* doubt that such bridges are possible in a practical sense, they are in no way doubt that resource management is needed and is possible, at least in theory.

What looks then to be a "debate" between advocates of sustainable development and Ludwig *et al* is really no debate at all. Both believe fundamentally in the need for improved resource management and both believe they know how that management should proceed. It is just that They Who Must Be Obeyed disagree over the rules and means to that management. There is a palpable sense in which not only is sustainable development debate an artificial negativity that reinforces the domination of the techno-managerial elite, but also that the complaints of the ecologists—we are too rarely consulted, we scientists cannot remain neutral—are themselves an artificial negativity on the part of ecologists who do not question the more basic issue: What, if any, alternatives are there to New Class resource managerialism? Meyer and Helfman reaffirm with obvious pride that ecology is "a subversive science,"[16] while in the same breadth making it clear that this "subversion" is nothing more than another kind of resource managerialism with global ambitions.

To see the artificial negativity at work, consider the statements of an ecologist and a Critical theorist. When the ecologist, C.S. Holling, writes,

> *The [key environmental and renewable resource] problems*
> *are essentially systems problems where aspects of behavior are*
> *complex and unpredictable... Therefore, interdisciplinary and*
> *integrated modes of inquiry are needed for understanding. And*

---

[13] Rubenstein (1993, 585-586).

[14] Salwasser (588).

[15] Levin (unnumbered page)

[16] Meyer and Helfman (571).

> *understanding (not complete explanation) is needed to inform*
> *policies...*[17]

his *Therefore* is a New Class "therefore," focused exclusively on the
management issue of making better policy. Critical theorists would follow the
very same first sentence in the quote with a very different "therefore," as in:
"*Therefore*, the pressure on localities to secede or otherwise decouple
themselves from system complexity is always immanent, if not imminent,
thereby working against any 'informed' (i.e., New Class) policies."[18]

Consider what an ecologist and Critical theorist would take as the core
environmental question. "How modern societies can live and prosper
'sustainably' on the planet is *the* greatest challenge facing humankind and
ecology is essential to addressing this challenge".[19] Put so baldly, one may
readily agree to the first half of the sentence, even if turning a blind eye to the
New Class thrust of the second half. Not so for Piccone. Rather, "the projects of
decentralization, de-bureaucratization, or re-empowering of individuals and
communities in order to allow a minimum of self-determination concerning life-
style, culture and identity, are some of the most important, if not *the* most
important political issue of today".[20]

The real threat to the New Class is that these local communities, if left on
their own, would use their "resources unsustainably" and that no one in the New
Class could stop them from doing so. Populism and communities surface in the
pages of *Science* and *Ecological Applications* only in two ways: either as the
greed and shortsightedness roundly criticized by all or as the "stakeholders" in a
"bottom-up rather than top-down" management process committed to "long-run
rather than short-run priorities" and operating according to "ecologically-sound
principles".[21] Human motivation must be instrumentalized, if it is to matter at
all. In New Class-speak, the management problem is to incorporate "the human
dimension" into an ecologically-informed decisionmaking.[22]

---

[17] Holling (554; highlights in the original).

[18] Even if this system complexity were not real, from a Critical theory perspective New
Class ecologists would have very likely invented the concept as a form of artificial
negativity to justify their continued work. In this view, major US problems are
complicated because they "arise," not to be solved, but to keep the New Class busy.

[19] Mangel *et al* (573).

[20] Piccone (1987/88, 7-8).

[21] See, for example, Mooney and Sala, Mangel *et al*, and Ehrlich and Daily.

[22] See, for example, Meyer and Helfman (570). "Include human motivation by developing
linkages with the social sciences, particularly economics, to develop a comprehensive
transdisciplinary synthesis," is how Costanza (580) puts what is a wonderful, albeit
unintentional, illustration of New Class artificial negativity at work—as if a major

In very much the same way, the *bête noire* of sustainable development proponents—"unregulated growth" or "unfettered resource exploitation"[23]—is itself a smokescreen of artificial negativity obscuring the organic. Here the artificial negativity is all that white noise about economic growth and resource exploitation as The Only Problem, where all the while the real message being told lies in those adjectives "unfettered" and "unregulated," the markers for all that is unmanageable and outside the control of anyone in the New Class. The notion that people, in their own communities and through populist movements, do what they want with and to their "environment," regardless of what transnational corporations and the environmental worldwatchers dictate, threatens both the New Class advocates and detractors of sustainable development. Heterogeneity *is* here to stay, and that's *the* management problem for the New Class: "Because the natural world is rich in patches, unsustainable use can continue for long enough for humans to assume it can be permanent...," bemoans one ecologist.[24]

In organic negativity, the New Class has no control over those communities which on the basis of their own culture and traditions "overgraze," "overutilize," and "overexploit their resources." Charges of greed and shortsightedness, like the charges of encouraging nationalist strife, are New Class ways of silencing populism without a hearing. What one *Ecological Applications* respondent calls "strong cultural beliefs" and another "strong tribal structure"[25] are okey-dokey for the New Class—as long as they lead to "sound" resource management. When they don't promote management, the practices are said to run counter to "common sense" and, heaven forfend, "the scientific method".[26]

---

problem with resource managerialism is the need to link human motivation to the social sciences!

[23] Respectively, Mangel *et al* (574); Meyer and Helfman (569). There is one sense in which sustainable systems and organic communities are similar, i.e., some would describe each as "self-correcting," though in fundamentally different ways For a few, the appeal of sustainable development is the promise that human systems can mimic ecosystems, in that they both can be self-correcting, i.e., self-sustaining. "Participants," to one recent conference on ecosystem management, "focused on the need to develop solutions that enhanced the flexibility of the system itself to respond to natural fluctuations and 'self-correct'" (California Urban Water Agencies, "Ecosystem Management: The Bay-Delta Solution," Final Report (July 1995), p. ES-2. Indeed, it is fairly easy to see in such comments an appeal to systems that need only a minimum of human interventions, New Class or otherwise (for more, see Chapter 4). The "corrections" put into place through organic communities and populism, of course, are nothing without those human interventions.

[24] Lee (561).

[25] Respectively, Mangel (573); Meyer and Helfman (569).

[26] See, for example, Zedler (578); Costanza (580).

Read this way, the claims New Class ecologists make—in nature, everything is connected; nature needs better managing—take on an anti-populist, even persecutory insinuation. According to ecologists, a much more holistic, global systems perspective to development is needed, one that recognizes the high interdependence and connection between all things.[27] From outside the New Class perspective, such assertions look decidedly imperialistic in their management aspirations, colonizing the lifeworld with globalizing New Class claims to panoptic expertise.

Look too at that New Class mantra, "need." While the language of needs is difficult to avoid, even in a critique such as this chapter, it is invariably New Class elites who talk in terms of "needs" that have to be filled and their special expertise in filling them. Needs assessment is the worst kind of artificial negativity peddled by the New Class, whereby people are led to believe that they are all alike in having "basic human needs," that there are always more needs than can be met, and that only the New Class can help them to meet those needs that would not be met otherwise. The techno-managerial elite alone has the diagnostic credentials to manage the therapeutic society.[28] And what better way to create needs in need of the New Class than by manufacturing them through a global, holistic systems approach which all but guarantees that even where local needs are met, global ones aren't. Development can only be a local success but always a system failure, when parsed through an insistence that the local provides little, if any leverage, in correcting what the New Class defines as "our" global problems.[29]

Populism, on the other hand, rejects all this interconnection and complexity and New Class claims to needs assessment. People don't "need" organic negativity; they *are* organic negativity. In organic negativity, people have abilities they manage on their own; in artificial negativity, people have needs that have to be met through outside managers. In populism, it is culture, tradition and particularity that matter; in the New Class, the only gurus are management gurus.

In this way, statements in *Ecological Applications* like "Halt population growth," "Reduce per capita consumption," and "Any cause is a lost cause without a reduction in population" become, from a populist perspective, nothing more than elitist hectoring and harassment in their ambition to homogenize heterogeneous peoples into a mass population, universalize their problems into the staccato "Reduce! Reduce! Reduce!", and abstractly individualize people into all manner of per capita's. Organic negativity, to repeat, will have none of this. It is only at the community level or in a populist movement, where abilities are defined locally and vary widely, do you really understand just how conformist and condescending are recommendations like "halt" and "reduce".[30]

---

[27] See Meyer and Helfman (570); Slobodkin (572).

[28] Slobodkin (572).

[29] Roe (1994b).

[30] Robert Solow, the Nobel economist, captures this persecutory tone when, in his discussion of sustainable development, he complains that "there is something faintly

Halt population growth or Reduce per capita consumption is not nor could ever be the answer for really existing communities.

By this point, the environmentally-conscious reader will be climbing the proverbial wall—ecologists and sustainable development advocates as "persecutory"? "anti-populist"!? "imperialist"??!! The reader fulminates: "Do you really believe in this conspiracy theory? They're the good guys, after all! Just what do you propose for a world that is destroying the environment all around us..."

The above critique is, to repeat, largely functionalist, rarely intentionalist. Of course, lawyers from a bureaucratic EPA when confronting lawyers from an equally bureaucratic NRDC mediated by an even more bureaucratic court system rarely see themselves in a conspiracy to keep the New Class bureaucracy in power. Of course, ecologists are not deliberately conspiring with decisionmakers who have excluded them from the decisionmaking loop.[31] Of course, applied scientists often have problems with those who preach the gospel according to basic research only.[32] In each case, we might well expect the principals to see themselves in adversarial terms.

Yet in each case, that polarization is artificial in functional terms, because in each case the underlying game that is being played—resource managerialism in a transnational register—is never in question among the "adversaries" concerned. What is not in doubt is a resource managerialism whose principles are that resources can be managed by experts who know what the real needs are, that this expert management matters for decisionmaking, and that the experts involved should be grounded in science and engineering. Except for Luke's critique, what has been missing in the sustainable development debate so far (and not just in the pages of *Science* or *Ecological Applications*) has been the populist critique of resource managerialism itself, namely: Management categories, like "resources" and "the environment,"[33] are an ultimately self-serving and reinforcing creation of experts; even where such categories have real referents, communities do not "need" outside experts to manage the resources; moreover, expert management does not matter as much as local control over these "resources" or "the environment," and certainly there is no reason to privilege or elevate science and engineers in *any* of that local control.

A populist critique of the "debate" between sustainable development and its detractors, like Ludwig, Hilborn and Walters, insists that what is really going in this so-called controversy is New Class experts jockeying among themselves in claiming rights to stewardship over land and resources they do not own. By

---

phony about deep concern for the future combined with callousness about the state of the world today" (quoted in Levin, unnumbered page).

[31] See, for example, Pitelka and Pitelka, and Rubenstein. It should go without saying that some ecologists see the "technocratic ideas" that motivate sustainable development (e.g., Lee, 560).

[32] For example, Salwasser (588).

[33] Mangel *et. al.*, for example, equate ecosystems with "renewable resources" (573).

generating and participating in such a debate, technical experts and managers assert rights as equal (instrumentalized) "stakeholders" in the land and resources they say are not being managed sustainably. Working on the assumption that those who sustain resources are the best stewards of those resources, the experts make a twofold claim, namely, not only are insiders, specifically local residents and communities, not stewarding their resources, but those who really know how to sustain those resources are outsiders, specifically the experts and professionally-trained resource managers. Accordingly, so this argument goes, local people are in need of the stewardship of systems-oriented techno-managerial elites, be they experts in governments, international donor agencies, or transnational nongovernmental organizations.[34]

"Wait a minute," interrupts the reader again," it's precisely because of local actions that we need global remedies. It's because people are polluting and destroying resources locally and regionally that we have such problems as climate change requiring global remedies now. All you're recommending is more of the same that got us in the environmental mess we are today..." If you believe that, then you have raised an empirically and logically prior question that must first be answered: What reason do you have to believe that if you try—and fail—locally, regionally, and nationally to reduce, e.g., CO_2 emissions, you will have any more success globally? Do you really believe, after four decades of trying to transfer technology and institutional arrangements from parts of the developed to developing world, that you will have fewer failures when transferring even more complex technologies and arrangements across altogether wider tracts of the planet to address your putatively global problems like climate change? We already have considerable grounds for reducing CO_2 emissions in the US—namely, we should be doing so for a variety of other reasons—without waiting to settle the global technology transfer problem or the issue of whether or not global warming is actually occurring.[35]

Indeed, the opposite tack is more plausible. If communities knew they only had access to finite resources falling within their territory or by arrangement with other communities, each community would then be under more pressure to manage them better—assuming, of course, they even see their world divided up into categories like resources and the environment. If they don't manage them effectively, no one can—and it is precisely that conclusion which has motivated the organic negativity of populist movements. As Piccone notes, the populism at issue here is "federal populism." Communities whose organic negativity lead to expansionism on their part, e.g., having exhausted their land, they now rise up to seize their neighbors', would be restrained by a loose confederation of the heterogeneous communities, which would act in concert only under such or similar provocation.[36]

---

[34] For more, see Roe (1995).

[35] Global warming is discussed in Roe (1994b).

[36] Piccone (1991); D'Amico and Piccone (1992).

CONCLUSION

The four questions that started this chapter can now be answered. From a Critical theory perspective,

> *What is sustainable development?* It is a New Class version of resource managerialism that functionally serves to globalize and perpetuate the techno-managerial elite's control over everyday life. While opposed to standard economic growth prescriptions, sustainable development is really artificial negativity that leaves untouched the issue of whether or not the New Class resource managerialism is appropriate at all.

> *Why is it an issue?* Ultimately it is an issue because the New Class' older versions of resource managerialism, such as economic growth and transnational exploitation, have come under increasing criticism, complexity and uncertainty, so that they now must be augmented and/or replaced by newer, aspiring versions of resource managerialism under the expert control of techno-managerial elites.

> *Ideally, what needs to be done?* Nothing "needs" to be done. Ideally, organic negativity would eliminate the "need" for New Class domination, in particular resource managerialism and its progeny, including sustainable development.

> *But practically what can actually be done?* Protect really existing communities which seek control of their own territories. Promote a loose federal arrangement predicated on heterogeneous communities, each with the right to secede from this confederation but having no right to usurp others. Don't be deluded into thinking in terms of the Left or the Right, as they both operate according to New Class principles. Give populism a chance.

ADDENDUM: IMPLICATIONS FOR CRITICAL THEORY

Unlike the other three approaches discussed in Part II of *Taking Complexity Seriously*, Critical theory is directly affected by what course the sustainable development debate takes. To see this, return to Luke's analysis of sustainable development. Luke—who is certainly not alone in his view—believes there is nothing like an "unmanaged" or "untouched and totally wild" nature out there. Human beings have interfered and will continue to do so in a "Nature" that is better thought as constructed rather than pre-existing. Even if the interference were one simply of fencing off a so-called "wilderness area" so that it would remain untouched by human hands, that very act of fencing off and preserving the area is itself no less a kind of management than managing the area otherwise.

Yet if that is true, then does the same hold for the relationship between federalism and the organic communities those federal arrangements are meant to protect? If the role of federalism is, as Piccone has insisted on a number of occasions,[37] to *protect* really existing communities from the encroachment of New Class outsiders, then is not that very act a kind of management no less interventionist than what is going on today under New Class resource managerialism? More to the point, unlike wilderness areas totally devoid of human beings, these protected communities would be populated—and quite emphatically so in the *Telos* model—by really existing people whose many activities would include management of, if not resources, certainly their territories.[38]

Conversely, if it is the case that strong communities give rise to and remain strong under a minimalist federalist management umbrella, then the implication is that true wilderness areas can and do remain on this planet, even where the manager's strategy is one of hands-off management of these "resources," whether out of respect for and in awe of that resource or through the resource's physical inaccessibility and remoteness to current human intervention.

On the other side, if all management is New Class management, then it is difficult to see how protected communities are any less "managed" by the New Class than penetrated ones, whether out of respect for and in awe of organic communities or through the inaccessibility and remoteness of organic communities to current New Class intervention. If the latter is true, then Luke's observation that areas protected by the likes of The Nature Conservancy are really Nature's cemetery (i.e., a memorial of a Nature long dead), then too, strong communities, even under a loose federal structure, run the risk of being the tombstones for an organic negativity and populism equally dead in the wider scheme of modernity.

---

[37] *Ibid.*

[38] See Piccone (1991, 38ff) on the importance of territory to organically constituted communities. For a criticism of Piccone's notion of community, see Fortmann and Roe (1993). For a very Picconesque argument about the difference between community and management, but from a different perspective, see Post (1995, 3-15).

4 SUSTAINABLE DEVELOPMENT AND THE LOCAL JUSTICE FRAMEWORK

What is sustainable development? Why is it an issue? Ideally, what needs to be done? and *Practically, what can be done?* are answered by parsing the questions through the local justice framework of the political philosopher and theorist, Jon Elster.

THE LOCAL JUSTICE FRAMEWORK

Jon Elster takes great pains to show that his is not a theory but a framework. In fact, a theory of local justice is really not possible in Elster's view, as the allocation systems that interest him—immigration, layoffs, college admissions, wartime rationing, and military service, among others—are too complex and messy to subsume under an all-embracing theory of justice such as John Rawl's or Robert Nozick's (e.g., Elster 1992, 15).

Elster's particular concern is to describe and explain the allocation of scarce—mostly indivisible and heterogeneous (Ibid, 23-24, 26)—goods and burdens made in-kind and largely outside markets by specific autonomous institutions to specific individuals. In this way:[1]

> *Local justice can be contrasted with global justice. Roughly speaking, globally redistributive policies are characterized by three features. First, they are designed centrally, at the level of the national government. Second, they are intended to compensate people for various sorts of bad luck, resulting from the possession of "morally arbitrary properties." Third, they typically take the form of cash transfers. Principles of local justice differ on all three counts. They are designed by relatively autonomous institutions which, although they may be constrained by guidelines laid down by the center, have some autonomy to*

---

[1] Jon Elster (1992, 4).

design and implement their preferred scheme. Also, they are not compensatory, or only partially so. A scheme for allocating scarce medical resources may compensate patients for bad medical luck, but not for other kinds of bad luck (including the bad luck of being turned down for another scarce good). Finally, local justice concerns allocation in kind of goods (and burdens), not of money.

The semi-autonomous institutions are local in one or all of three senses: arena, country and locality. In the US, different arenas, such as organ transplantation, college admissions and job lay-offs, follow different principles: "*Need* is central in allocating organs for transplantation, *merit* in admitting students to college and *seniority* in selecting workers for layoffs" (Ibid, 2). Allocative principles vary by country as well: "In many European countries, need (as measured by number of family dependents) can be a factor in deciding which workers to lay off" (Ibid, 3). Lastly, allocative principles can vary by locality within the same country or arena, as with the case of local transplantation centers in the US (Op Cit).

Complexity in local justice systems comes not just from the fact that the goods are scarce, heterogeneous and in kind and that the allocation may well be local in multiple senses. Local justice systems vary because principles are tied to complex (and not always consistent) criteria, mechanisms, procedures, and schemes for allocation.[2] As important, the actors in local justice systems are multiple. *Politicians*, in Elster's terminology, make first-order decisions over how much to allocate; *allocative officers* make second-order decisions over how to distribute the amount to be allocated (these on-the-ground allocative institutions are the special province of local justice systems); and *recipients* make or respond to third-order decisions (particularly incentives and disincentives) that either affect the recipient's need for the scarce good being allocated or their likelihood of receiving it (Elster, 5, 139ff). These actors, in turn, are motivated by different principles. From Elster's perspective, first-order politicians are motivated primarily (albeit not exclusively) by efficiency concerns; second-order allocators by equity as well as efficiency; and third-order recipients by self-interest (Ibid, 180-182). Elster talks also about a fourth "actor," that of a diffuse public and public opinion, especially in the form of the media (Ibid, 5). For Elster, this hodgepodge of factors such as heterogeneity,

---

[2] "I shall use the term *principle* to designate any general conception of how the scarce good is to be allocated. Some principles refer to substantive properties of individuals, such as 'to each according to his need" or 'any close relative of a citizen is allowed to immigrate.' I shall refer to these properties as *criteria*. Other principles are embodied in *mechanisms*, which do not require individualized knowledge about the potential recipients. Equal division, lotteries, and queuing are prime examples. I shall refer to *procedures* as the operationalized version of a principle...I shall sometimes refer to an allocative *scheme* as the full set of procedures used to implement mixed principles" (Ibid, 62-63).

scarcity, localisms, and multiple actors subscribing to mixed principles and practices makes for compelling complexity in local justice systems.

Complexity, but not unintelligibility. Elster believes his classification of principles is fairly complete, for example.[3] Elster also describes a fairly coherent "commonsense conception of justice," which he believes is held by most first-order politicians, centering around the maximization of welfare, subject to a fixed minimum for individual welfare but excluding people who drop beneath that minimum through their own choice rather than through poverty or deprivation (more in a moment). Lastly, while variousness is characteristic of local justice on the ground, the processes whereby they arrive at this variability—through coalition building, bargaining and compromise, and historical accumulation and accretion (Ibid, 172-178)—-are found across cases.

Local justice systems have clear implications. Not only are they not designed to compensate for global injustices, local justice systems can actually lead to those injustices (Ibid, 133-134):

> *From childhood to old age, [the individual] encounters a* *succession of institutions, each of which has the power to give or deny him some scarce good. In some cases the cumulative impact of these decisions may be grossly unfair. We can easily imagine an individual who through sheer bad luck is chosen for all the necessary burdens and denied all the scarce goods, because in each case he is just below the cutoff point of selection. To my knowledge this source of injustice has not been recognized so far...Those who are entrusted with the task of allocating a scarce good rarely if ever evaluate recipients in the light of their past successes or failures in receiving other goods. Local justice is largely noncompensatory. There is no mechanism of redress across allocative spheres....*
>
> *For most people, events will turn out so that they can say to themselves, "You win some, you lose some." But by the nature of chance events, some individuals will miss every train: they are turned down for medical school, chosen by the draft lottery, laid off by the firm in a recession, and refused scarce medical resources; in addition, their spouse develops cancer, their stocks become worthless, and their neighborhood is chosen for a toxic waste dump. It is neither desirable nor possible to create a mechanism of redress to compensate all forms of cumulative bad luck. For one thing, the problems of moral hazard would be immense [i.e., if people knew they were going to be compensated for whatever happens to them, they could take more risks and*

---

[3] Elster (67). He itemizes and then discusses what he calls egalitarian principles (e.g., absolute equality), time-related principles (e.g., queuing and seniority), principles defined by status (e.g., age, gender and any number of other characteristics), principles defined by other properties (e.g. need and efficiency), mechanisms based on power (e.g., purchasing power), and mixed systems that combine the preceding principles (Ibid, Chapter 3).

> *thereby incur more harm]. For another, the machinery of administering redress for bad luck would be hopelessly complex and costly.*

Elster (134) hastens to add, "These objections do not, however, exclude all forms of inter-institutional compensation." "There is probably not much we can achieve along these lines," however.

THE FRAMEWORK'S REFORMULATION OF THE SUSTAINABLE DEVELOPMENT DEBATE

You might expect sustainable development to have everything to do with local justice systems. Sustainable development is fundamentally concerned with the allocation of scarce and heterogeneous goods and services; so is local justice. Sustainable development is a response to complexity; so is local justice. Sustainable development is nothing if not preoccupied with "goods" that are in-kind and cannot be reduced to money alone; so is local justice. Sustainable development is driven by concerns of efficiency and equity, particularly through second-order institutions charged with allocating these goods; so is local justice. What motivates sustainability and local justice, in short, appear to be very much the same.

Yet the discussions in *Ecological Applications* and Elster's *Local Justice* could not be more worlds apart. The focus of *Ecological Applications* debate on overpopulation and resource overutilization bears scant resemblance to Elster's discussion of related issues like immigration and CO_2 emissions. This gap has much more to it than just the fact that sustainability is based on biological considerations in ways that local justice systems are not.

The difference between the two approaches is best illustrated at the level of principles to which Elster and the *Ecological Applications* respondents appeal. For Ludwig *et al* (549):

> *Most principles of decision-making under uncertainty are simply common sense. We must consider a variety of plausible hypotheses about the world; consider a variety of possible strategies; favor actions that are robust to uncertainties; hedge; favor actions that are informative; probe and experiment; monitor results; update assessments and modify policy accordingly; and favor actions that are reversible.*

On the other hand, Elster (240) has his own common sense formulation of what he sees as the driving issues of justice and welfare:

> *The commonsense conception of welfare may, then, be stated in four propositions, each of which modifies its predecessor. (1) Maximize total welfare. (2) Deviate from that goal if necessary to ensure that all achieve a minimum of welfare. (3) Deviate from the requirement of a minimal level of welfare in the case of*

*persons who fall below it because of their own choices. (4)
Deviate from the principle of not supporting the persons identified
in (3) if their failure to plan ahead and react to incentives is due
to severe poverty and deprivation.*

The temptation is to reconcile the two sets of principles by insisting that the first is to be used whenever trying to realize the second. Justice, however, is by no means always an experiment, it may have little time for caution, while having everything to do with being irreversible. Indeed, there may be nothing more unjust, from Elster's common sense perspective, than the blanket application of "Halt population increase" and "Reduce per capita income" favored by respondents in *Ecological Applications.*

What, then, do layoffs, draft boards, kidney transplants, college admissions, child adoption, and the rest of Elster's local justice systems have to do with the sustainable development of *Ecological Applications*? To answer that question points to a major failing in much of the sustainable development debate, and not just in the pages of science journals. For the answer is, From the perspective of the local justice framework, sustainable development means little if local justice systems are not central to its imperative.

Several factors account for the disjunction between *Ecological Applications* and *Local Justice*. First, in a very real sense, a significant part of the sustainable development debate has been devoted to first-order decisions over whether to allocate resources, and if so, to whom. Several *Ecological Applications* respondents refer to politicians, resource managers, the public, media and the like (Hilborn and Ludwig, 550; Holling, 553; Ehrlich and Daily, 558), while giving only passing attention to second-order decisions affecting how those resources are allocated, once their levels and target populations have been identified. This is a profound omission in the *Ecological Applications* debate, not simply because the gap between first-order decisions and their second-order implementation is large, but because second-order actions clearly influence first-order decisions (Elster, Chapter 5). Most obviously, how we conserve resources today has substantial effects on how many resources are available in the future for allocation. Since second-order decisions are for Elster the special province of local justice systems, the relative silence of the *Ecological Applications* debate over these systems is not altogether surprising, as this is a debate preoccupied with first-order questions of getting "the big picture" right. Ecologists, for example, are much more comfortable recommending ecosystem management than detailing how it is to be done.

The debate's silence extends as well to third-order decisions and actions of the recipients of the scarce goods and burdens being allocated. If sustainable development is really about the end of greed as Lee (562) implies, then we must look at the incentives *and* disincentives put into place with the implementation of the kinds of sustainable development envisioned by the respondents in *Ecological Applications.* What perverse incentives and disincentives would be created, were one to adopt the kinds of positions outlined in that journal's debate? In what ways are the admonitions, "Halt population growth" and "Reduce per capita consumption," any less simplistic than the "Maximize sustained yield" so roundly criticized by that journal's respondents (e.g., see Salwasser, 588)?

Third-order decisions also clearly affect sustainable development. If people are wasting their time in order to avoid military service or get into medical school (e.g., by undertaking compulsory community work better undertaken by others in Elster's view), then this surely has an effect on the potential for sustainable development. The debate in *Ecological Applications* is not alone in having little, if anything, to say about the myriad of specific issues, such as admission to military service or medical schools, when it comes to realizing sustainable development.

All this points in one direction: Ecology may be complex all the way down, but sustainable development as currently discussed by ecologists is not. In *Ecological Applications*, sustainable development isn't an admixture of principles, criteria, procedures and schemes in the way Elster's local justice systems are. Nor could sustainable development be, if simply because local justice systems are designed to vary on the ground in ways that sustainable development presumably could not vary. Local justice systems are decentralized and uncoordinated, while it is clear from the *Ecological Applications* that sustainable development is not to operate in a decentralized and uncoordinated fashion. Local justice systems are a product of coalition building, bargaining and accretion, whereas sustainable development in the *Ecological Applications* debate is anything but that.

In fact, ecologists see in sustainable development the long overdue repudiation of all this coalition building, bargaining, accretion, and lack of coordination. Sustainable development is for many very much the rejection of a human discretion that occurs in the absence of binding guidelines and principles (see Elster, 63). What appeals to many of its advocates is the promise sustainable development holds out for producing self-sustaining or self-correcting systems (Mangel *et al*, 573)—that is, systems defined precisely by their minimum of human (mis)management and (in)discretion. Whatever sustainable development is in practice, it wants no more "local justice" discretionary power in the hands of those who have historically misused that discretion.

Sustainable development is not without its own set of principles or concern for equity. The set, though, is abbreviated from the perspective of the local justice framework. In the sustainable development debate of *Ecological Applications* there is one primary principle, that of the need for system sustainability (sometimes called system "health" or "integrity" and analogous in respects to Elster's welfare terminology[4]); and the equity of interest is primarily that of intergenerational equity (Levin, unnumbered page).[5] Presumably such

---

[4] Elster purportedly complete list of criteria does not include "ecosystem sustainability" or "health." Although he does talk about "health" (186), it is clear Elster's focus is on human beings only. As shall be seen shortly, within the primary principle of sustainability are other principles, including the so-called precautionary principle.

[5] Meyer and Helfman (569): "This concept of intergenerational equity is the backbone of sustainability."

concerns can also be found in some local justice systems, but, as described by Elster, these systems are denser and more textured in terms of principles, criteria, procedures and schemes than that system sustainability subscribed to in *Ecological Applications*. "Actual allocative systems can virtually never be reduced to a single principle," in Elster's view (103).

Part of the reason why local justice systems are more decentralized, uncoordinated and variegated than sustainable development stems from the fact that the former is about individuals and the local, while the latter is more about society and the global. Local justice is preoccupied with specifics, sustainability is very much about the general. As Robert Solow puts it, sustainable development "is a general obligation, not a specific one" (quoted in Levin, unnumbered pages). Perhaps the only time sustainability and justice are explicitly tied together in the *Ecological Applications* debate is at the global level, when Robert Costanza notes that the precautionary principle in sustainable development (i.e., regulators should take into account any environmental harm before acting) "has come to be seen by some as a basic normative principle of international environmental law" (580).

Rather than the specific and "local," virtually all the ink in *Ecological Applications* is devoted to a "scale" that has nowhere to go but up. Consequently, much more variability can be expected from local justice systems than from sustainable development, if simply because variability in the latter has been aggregated away into the general, the international, the global (for an exception, see Meyer and Helfman, 570). That said, it is disingenuous for respondents in *Ecological Applications* to insist, as they seem to, that we must think globally but act locally, without then explaining how this can or should be done at the second- and third-order decision levels of implementation and operation.

From a local justice perspective, a globalized sustainable development is rooted in ironies, not least of which is the fact that what makes for broad, coherent approaches like sustainable development—national culture, professional norms, and political regulation (Elster, 138)—are exactly the things a generalized sustainable development is against.

What, though, about the relationship between sustainability and intergenerational equity? In the Elster terminology, the need for sustainability is both a level and an increment need, which are often conflated.[6] For example, we are told to save what remaining biodiversity there is, without telling us which is the more important need—to concentrate first on saving as many species as we can (the increment) or on those who are under the greatest threat (level) (see Ibid, 93). Saving from extinction and maximizing the number alive for the future may well be mutually exclusive tasks, when we cannot do all that is "needed."

Part of the conflation is that much of the sustainable development debate, and not just in the pages of the *Ecological Applications*, subscribes to an all-embracing diagnostic needs assessment approach:

---

[6] Elster spends considerable time on describing need in terms of the level criterion of welfare (84, 170).

Ecology does not have the simple problem set or rigorous theory of astronomy, but it is comparable to medicine... Particularly in cases involving illnesses in one patient, medicine shares with ecology extremely limited controls and replicates. Each medical diagnosis is a miniature research problem into possibly unique phenomena. The empirical and theoretical background of the physician increases his or her competence and speed at reaching a diagnosis and a treatment protocol. Similarly, an applied ecologist of broad competence and background can develop a solution to a specific problem more quickly and with a higher probability of success than one of lower competence.
(Slobodkin, 572)

What such a medical analogy misses are those other kinds of "needs" assessments more consonant with Elster's focus on coalition building, bargaining and accretion. There are a variety of other ways for assessing what needs are in addition to the diagnostic, including what Rothe (1991) calls the democratic view. In the latter case, need arises not because something harmful has been diagnosed, but because those in the majority of some reference group desire something else than what is currently the case. When more and more ecologists say there is "need" for ecosystems that mimic presettlement conditions, they are not diagnosing need as much as building consensus for one. The pretense that what ecologists are doing in the sustainable development debate is diagnosis has come close to betraying sustainable development as just a one-dimensional need, i.e., there is nothing more urgent than saving the planet, period. As Elster reminds us, local justice systems ensure that there are many other urgent needs, period.

Turn now to the issue of intergenerational equity. Sustainable development advocates believe that the concern for intergenerational equity follows directly from sustainable development's definition as using resources today in ways that leave options open for the future's use of these resources. There is, though, an implicit local justice dimension in this. Elster notes that, in local justice situations of resource conflict, equality "is often the only focal point for the resolution of conflict," even when there is no consensus over what "equality" is fair and even when envy motivates the drive to equality (Elster, 70, 227). Consequently, it is not surprising that a sustainable development which focuses on resolving resource conflicts would also focus on equality, in our case intergenerational equity. Where conflict has to be settled, concerns for equality show up as the arbiter of last resort. Here the insistence of last resort is that all future generations are equal in some important respect.

Are they all equal? Elster points out that lotteries are one way local justice systems allocate goods by giving people an equal chance of getting them: "Lotteries respect the spirit of equality, while avoiding the conclusion that a good must be wasted if it cannot be given to everyone" (72). If there are resources today which cannot be conserved for everyone all time, does it then make sense to have a lottery to pick the generation(s) for which we will try to conserve the resource? What sense does it make to conserve oil, for example, for the 30th generation from now, if that were the generation we pick in the "sustainability lottery"? One does not need to be a "posterity-be-damned" advocate to wonder in what way intergenerational equity has any rationale that holds across persistently variable local systems.[7]

Such are the major criticisms of the *Ecological Applications* debate from a local justice framework. The question now becomes: What would a sustainable development look like if made consistent with Elster's local justice framework?

The starting point[8] of such a reformulation would have to be where systems of local justice and sustainability overlap most clearly: They both start in complexity. Elster quotes with approval the observation that complexity is itself a form of defacto allocation (Ibid, 115). Ecosystems are systems of complex allocation, just as are human systems—not intentionally, perhaps not even functionally, but as a matter of after-the-fact. It is also quite clear just what kind of complexity sustainable development must be from a local justice perspective: It cannot be just one principle, but arrays of principles, criteria, procedures, mechanisms and the like that vary across space and time. It must be focused on levels and increments of welfare at the same time. It cannot be just about first-order decisions, but also about second- and third-order decisions. It must be complex all the way down.

The temptation is to get specific about what these locally varying factors should be or are in practice. The bigger issue raised earlier, though, has first to be addressed. How can development be sustainable, but still be uncoordinated and decentralized at the same time? More formally, how can you have what is ostensibly a globalized sustainable development that is everywhere localized and variable? Is that not a contradiction in terms?

More paradox than contradiction. Start with Elster's insight that local justice can lead to global injustice. From a local justice perspective, sustainable development poses the opposite problem. The global justice of sustainable development, like any other universalized approach to goodness, can lead to local injustices, when the former is implemented uniformly over an altogether differentiated landscape. For Elster, the gap between local justice and global injustice is largely a gap that cannot be filled. So too by extension is the gap between global justice and local injustice unbridgeable. You can't make the omelet of sustainability without breaking the eggs of local discretion.

Such disjunctions raise the specter of justice/injustice cycles at work in sustainable development as well as in other areas. To break arbitrarily into such a cycle: ...the more local injustice there is, the greater local pressure for more just systems. The more systems become locally just, the more global injustice can be expected. The more global injustice, the greater the global pressure for more just systems. But the more global justice there is, the greater the chances

---

[7] "Posterity-be-damned" advocates are less concerned with egalitarian principles of allocation than with time-related ones. We of the present matter more than future generations because we've spent longer time in the queue waiting for goods and services than those who have yet to be born or who have already died. For Elster on queues and other time-related principles of allocation, see his (73ff).

[8] The following assumes that sustainable development is more than what Elster calls the "norm of compassion" applied to non-human things, in this case nature and ecosystems (see 147).

for local injustice. And so on, where local injustice leads to local justice, local justice to global injustice, global injustice to global justice, and global justice to local injustice.[9]

To rephrase the cycle for the purposes of this chapter, the more unjust the global economy becomes, the more pressure there is for sustainable development strategies everywhere. The more sustainable development becomes a global phenomenon, the more locally unjust its uniform application would necessarily be. The more uniform the application, the more local pressure for suitably heterogeneous alternatives. But the more heterogeneous the applications of sustainable development on the ground, the greater the chance of global injustice arising from the decentralization and lack of coordination on the ground. In this way, just as it is not possible for local justice systems to compensate for the global injustices they create, so too it is not possible for global justice systems to compensate for the local injustices they create. Obviously, the speed with which such complex cycles operate depends on the degree to which the situations of interest are tightly and/or loosely coupled, i.e., Cells 2, 4 or a mix of both.

From this wider perspective, "sustainable development" can be reconceived as a justice/injustice cycle, and not just equated to one of its moments as is commonly done by most sustainable development advocates. The local justice framework insists that the local may well subvert globalisms like that sustainable development. In this way, sustainable development must be seen as not ending when unjust global systems become more just, but rather as continuing through a set of iterations whose moments include a *rejection* of an overly globalized sustainable development. What keeps the cycle one of sustainable development is not its "success" at one particular point, but its constantly coming back to that insistence that resources should be used today in ways that keep options open for tomorrow, whether locally or globally.

The implications of sustainable development as a justice/injustice cycle range from the banal observation that what is sustainable locally may unsustainable globally to the rather more unpalatable recognition that global sustainable development may have to be locally unjust and that some local justice systems can be just by *always* having priorities other than sustainable development.

CONCLUSION

We are now in a position to answer the four questions that started this chapter. From a local justice perspective,

> *What is sustainable development?* At best, it is a justice/injustice cycle that recurrently comes back to the notion that we should manage resources today in ways that leave open future options for their management.

---

[9] While Elster does not specifically address such cycles, he recognizes the to-ing and fro-ing of such pressures (e.g., 183).

Why is it an issue? Ultimately it is an issue because local justice systems are fairly complex in practice, which means in part that the justice/injustice cycle called sustainable development will inevitably involve elements of injustice throughout its operation.

Ideally, what needs to be done? Ideally, the complexity of local justice systems should be reduced, the homogeneity of sustainable development strategies complexified, or both.

But practically what can actually be done? Be prepared to accommodate the reality that local justice systems are inevitable and, as such, they will *invariably* alter sustainable development strategies on the ground.

5 TRIANGULATING ON SUSTAINABLE DEVELOPMENT

We are now in a position to answer the four questions motivating the chapters of Part II: *What is sustainable development? Why is it an issue? Ideally, what needs to be done?* and *Practically, what can actually be done?* The four approaches—Girardian economics, cultural theory, Critical theory, and the local justice framework—are ready to be used to triangulate on the answers. The implications of the triangulation for understanding and practicing sustainable development are as novel as they are considerable.

This concluding chapter has three sections. The first describes triangulation, the second answers the four questions, and the third provides final reflections on the exercise. Part III is devoted to drawing out some practical implications of the triangulation.

TRIANGULATION

Triangulation has been around for some time. As noted in Part I, triangulation is conventionally treated in the social sciences as the use of multiple methods, databases, theories, disciplines and/or investigators to study the same object, event or phenomenon (e.g., Denzin 1970).[1] Examples range from everyday ones of cross-checking statements from one source with other sources to more formal ones, such as relying on different (e.g., quantitative and/or qualitative) research methods to investigate the same issue or different theories to converge on a set of testable hypotheses (see Moris and Copestake 1993). The most popular form of triangulation has been the use of multiple—the "tri" need not refer to just three—methods (e.g., Brewer and Hunter 1989), and methodological triangulation figures prominently across a number of applied fields, including policy analysis, anthropology, marketing, investigative journalism, and participatory rural appraisal (Chambers 1992; Moris and Copestake 1993).

Triangulation is appealed to for several reasons. It is most appropriate when issue complexity and uncertainty is high, such as in sustainable development.

---

[1] Triangulation is a type of interdisciplinarity. Appendix 1 locates triangulation within the other interdisciplinary models.

"Many social scientists now assume a highly contingent world in which few relationships are so dependable that they hold across a wide variety of persons, settings, and times," in the words of the well-known methodologist, Thomas D. Cook (1985, 41). In such situations, "when it is not clear which of several options for question generation or method choice is 'correct,' all of them should be selected so as to 'triangulate' on the most useful or the most likely to be true" (Ibid, 38). (We will return to the question of "true" in a moment.) Denzin (1970, 26) puts the matter forcefully: "[B]ecause each method reveals different aspects of empirical reality, multiple methods of observations must be employed. This is termed triangulation and I now offer as a final methodological rule the principle that multiple methods must be used in every investigation, since no method is ever free of rival causal factors". Again, the methods can be quantitative *and* qualitative, holistic *and* reductionistic, postivist *and* post-postivist.

Triangulation is especially helpful in identifying and compensating for biases and limitations in any one instrument, be that "instrument" a theory, method, investigator or other source of information. "Triangulation is an important weapon in exposing bias," Moris and Copestake (1993, 48) remind us: "Obtaining a second expert opinion is an obvious example; but identifying all key stakeholders with an interest in a given matter and interviewing several with widely differing positions is even better" (see also Cook, 1985). Detecting bias is fundamental to complexity analysis and management (and not just management for sustainable development), because, as discussed in Part I, reducing or correcting for bias is one of the few things that researchers, managers and analysts can actually do when addressing highly complex and uncertain issues (Sorensen 1992, 262).

Just what is "complexity"? As we saw in Part I, complexity has three operational elements (Demchak 1991, 28): number of components in the system of interest, differentiation among those components, and interdependence between them. In this way, complexity of any system—be it ecological, social, technical, other—can be increased simply by adding more parts to it, assigning more tasks (functions, roles) to those parts, and/or ensuring these parts are interconnected in more ways than before. Thus, an important—some say the most important—feature of complexity is that surprise discussed in Part I, since complexity means we do not know the full causality of how things work in the system under study (Ibid, 17; Perrow 1984). By this point, it should be clear that these elements of complexity and its principal feature surprise are found at the heart of the *Ecological Applications* debate over sustainable development.

The virtue of triangulation, in the face of such complexity, is that it builds confidence we are on the track of something relevant or useful for our analysis. "When the findings of different methods agree, we are more confident," as Brewer and Hunter (1989, 17) conclude. Cook (1985, 46) is more explicit:

> ...when results are demonstrably stable across populations,
> settings, and times, external validity is enhanced...When results
> are stable across multiple potential threats to causal inference,
> internal validity is enhanced. And when interpretation of the
> meaning of relationships in theoretical and value terms is common

> *across a wide variety of perspectives, objectivity—defined as*
> *intersubjective verifiability—is enhanced.*

The point, however, is not that convergence in triangulation enables us to get closer to the "truth." Better to say, the use of multiple instruments, each with its own flaws, confirms the complexity we are analyzing. Instead of truth, what we get out of triangulation is confidence, i.e., convergence across multiple instruments enables us to be more assured we have in that convergence a point of departure which we feel is worth pursuing further. Rather than finding what is true, we triangulate because we don't know what is true, that is, we don't know the full causality at work. Which, of course, is what is meant by saying the issue is complex.

The failure to triangulate also provides useful information. "When data patterns do not converge across multiple measures, methods, populations, settings, times, and the like, the search begins to identify the contingencies controlling the relationship" (Ibid, 46-47). What we are studying may not be in any sense generalizable and triangulation is one of the few ways we are able to confirm what is a case in its own right. Much of sustainable development is non-generalizable in just this way, as shall be seen in a moment.

Triangulation poses problems, however. First, it can be time-consuming and expensive: "[R]estrictions of time and money make it impossible to employ multiple observers, multiple methods, and multiple data sources," Denzin (1970, 312) cautions. It took me over a year to write the preceding four chapters, as each required full immersion in the theory and its application. Second, successful triangulation is neither a necessary nor sufficient condition for effective policymaking in a Cells 2 and 4 world—though, as we saw in Part I, the same holds for any analysis in the face of pervasive complexity and uncertainty. Another problem with triangulation is, ironically, bias. While triangulation helps correct for the bias in any one instrument, care has to be taken to ensure that the biases are not in the same direction (see Cook 1985, for a detailed discussion of this issue). No matter how many cross-checking questions in a survey, they cannot correct for the fact that the interviewer is, e.g., white, male, middle-class and asks questions only in English.

The instruments used in triangulation should thus be as radically different (indeed, orthogonal) as possible: "Convergent findings are compelling only if it can be demonstrated empirically that when the methods err, they typically err in the opposite ways" (Brewer and Hunter 1989, 18). This demand for divergent instruments is the primary reason for choosing the four approaches used in the analysis of the sustainable development debate.[2] Since "management" is so

---

[2] The other two reasons for choosing the approaches are, to repeat: (1) each takes complexity and uncertainty seriously, i.e., each represents at some fundamental level a theory uncertainty and (2) each defamiliarizes the problem of sustainable development in fresh ways. Each approach has the added virtue of being an exceptionally potent theory or framework about which readers should become acquainted, and each raises hard issues that many in sustainable development would rather avoid or are silent on, such as the arbitrariness and injustice of many popular sustainable development pronouncements at the global level (more in a moment).

crucial to the definition and discussion of sustainable development (and not just in the pages of *Ecological Applications*), the approaches were selected precisely because they differed so dramatically on that management dimension. In a rough way, they mark off the major directions of a 360 degree compass of different approaches to resource management. As should be clear by this point, Critical theory sees the techno-managerial elite as dominant and ultimately corrosive of local community and culture (x); in contrast, the local justice framework finds such techno-managers in precisely these local communities and cultures, though not in all of them (some not-x); cultural theory sees such hierarchist techno-managers—local, global, wherever—as but one of a handful of basic cultures (both x and all not-x); while Girardian economics sees no such cultural/managerial distinctions persisting through time, as they will all eventually be washed way in another crisis of undifferentiation (neither x nor not-x).[3] Since the approaches have been chosen precisely because they are approximately orthogonal to each other, convergence in triangulation does *not* mean common ground has been identified, for how can you have common ground between polar opposites? Just what is being converged on will become clear by the end of this chapter.

For present purposes, the fact that each approach is diametrically different from the other three means each approach effectively functions as a critique of the others. Their status as critique explains why so little space has been devoted in these pages to criticizing the assumptions and premises of the approaches. Each, of course, is fairly easy to criticize. *All* prices are determined through mimetic polarization? *Only* four or five basic cultures? Politicians are *primarily* concerned with issues of *efficiency*? No New Class managers in *any* populist movements? Such questions, however, are misleading because, in the last analysis, they seek to determine which approach is right, when the point of triangulation is that all approaches are "flawed" in fundamental respects. I am always asked which of the four approaches I think is the best, and each time I answer that is not what triangulation is about. Triangulation in these pages is about convergence and confidence, not "truth," at least when the subject is as complex as sustainable development.

In triangulation, you do not need to devote much time exposing the problems in any one approach, as use of the others is itself the explicit recognition that the approach in question has serious shortcomings if used on its own. Critical theory faults the local justice framework because the latter misses communities as an example of local justice cross-compensation.[4] Cultural theory faults Critical theory by insisting there are more cultures than just hierarchists and egalitarians (the New Class techno-managerial elite and organic negativity, respectively). Girardianism faults cultural theory by insisting that all cultures

---

[3] The four approaches are not, however, orthogonal on all dimensions, e.g., Girard, Elster and Thompson *et al* all give attention to the importance of envy in human behavior.

[4] In an organic community and contra Elster, bad luck in one sphere can often be compensated by goods and services from another, as when the child who takes care of her/his sick parents is the last to be laid off from a town job.

and their conventions are inevitably corroded away in cascades of mimetic contagion. And the local justice framework faults Girardianism by insisting that local differentiation is much more complicated than Girardians acknowledge.

Because such approaches differ so greatly, they cannot all be the case at the same time for the situation under discussion. In triangulation you are using approaches you know to be wrong—what a difficult position for Cell 1 methodologists to accept!—but you don't know which ones are wrong or the way in which they are wrong because the problem at hand is causally so complex. As such and in contrast to Cell 1 research methods, triangulation confirms upfront the complexity of that which is being triangulated on. Thus, if and when such vastly different theories do converge, that convergence becomes all the more potent a basis for confidence and assurance that we are on the track of something useful for further analysis. Once again, there are no guarantees here that triangulation will lead to convergence or that this convergence, if realized, will always be useful.

A further tenet of triangulation is that the object of analysis should remain the same for all instruments used in the triangulation: "The first and most obvious problem [in triangulation] is locating a common unit of observation against which various theories can be applied. If each theory proposes a different set of propositions, then it follows that each also designates a different set of relevant data. How," Denzin (1970, 311) continues, "may observers be assured so that any test of these perspectives is valid and comparable?" His answer: "The only solution is to select one common data base and simply force the theories to be applied to those data" (Op Cit). In our case study, the common data base is that debate in the pages of *Ecological Applications*. Each of the four approaches in the triangulation has relied on the same published record and, often, the very same statements.

A special problem with triangulation deserves attention. Brewer and Hunter (1989, 18) point out that triangulation invites comparison and evaluation of the various instruments used, i.e., "multiple measurement offers the chance to assess each method's validity in light of other methods." Denzin (1970, 304) goes further and recognizes the temptation to select the best from each instrument (theory or method) and combine these elements into a composite instrument. For example, in a triangulation that uses different theories, the investigator has the chance "to develop his own propositions and theory...that only pertain to the data under analysis."

Nor is it difficult to image such a composite theory from our four approaches. When the head of the IMF complains that his institution is being made a "scapegoat" for unsuccessful reforms undertaken by national governments (Friedman 1994, A1, C6), he can be interpreted as saying that a techno-managerial elite of outside, hierarchically-oriented experts could have "saved" these countries by managing "their" resources better and more fairly, even on a local scale.

Yet, while one may not be able to resist comparing instruments, one must be very chary of making a composite of them. Remember, the theories or methods used in triangulation have been chosen because they are as different as possible. Triangulation is not about coming up with a new theory or methodology, but rather with the convergence, if any, of existing theories and methods on points of departure that the analyst feels confident in pursuing.

Theories, for example, can be combined to more fully explain the contingencies of a specific situation—as was just done in the IMF example—but these contingencies are by definition not general. Where the representativeness of the initial situation is not established, the alternative—in this case, triangulation—is to appeal to theories and methods that were designed to be general enough to explain these and other situations, albeit never to explain any one situation fully.

Denzin's methodological rule is NOT that, when faced with a complex event to explain, theories and methods should be combined to suit the situation. The rule is more stringent: When faced with situations of high complexity, you triangulate using different theories/methodologies. In our case on sustainable development, this means using the different perspectives of Girardian economics, cultural theory, Critical theory and the local justice framework to converge on a new point (or points) of departure. It does not mean making some new composite theory by mixing a bit of Girard, with a little of Elster, more of Piccone, and topped off with the best of Douglas and Wildavsky. Such a mix allows you to delude yourself into believing you increase certainty simply by constructing a composite instrument that seeks to better explain local contingencies; triangulation, in contrast, insists that you first ratify pre-existing uncertainty by acknowledging these local contingencies can never really be explained through triangulation of theories or methods that are by definition simultaneously wrong or otherwise deficient in important and fundamental respects. It is exactly in this way that triangulation is a kind of Cell 2 and 4 thinking for a Cell 2 and 4 policy world. Unfortunately, in this way as well, triangulation means loss of a great deal of "insights" that were identified by one, though not all, of the approaches concerned.[5]

Why, then, triangulate? Put yourself in the shoes of most policymakers or managers. Suppose you come upon a new theory or method, which, when applied to the case at hand, leads to a very different conclusion than the four used in the triangulation below. Is there any doubt what the average policymaker or manager would prefer—a situation where most of the very different theories or methods converge on what to do generally or a tailor-made instrument that explains one situation specifically?

Let us now turn to the results of triangulating on sustainable development using the analyses and findings of the preceding four chapters.

A TRIANGULATION ON SUSTAINABLE DEVELOPMENT

First to summarize the results of the following triangulation:

> *What is sustainable development?* Although much more complicated, it is about using resources today in ways that keep open options for their use in the future.

---

[5] The preceding chapters are rich in insights that do not reappear below. Unfortunately, they are not "information" until they are triangulated upon from the other perspectives.

Why is it an issue? It is an issue not just because we know few, if any, ways to keep options open for the future, but also because of a host of other factors too numerous to name.

Ideally, what needs to be done? Ultimately, what needs to be done is to reduce the complexity that gives rise to the demands for sustainable development in the first place.

But practically, what can actually be done? Differentiate! Beware the persecutory language of talking certainly about uncertainty. Acknowledge the unmanageable. Search for feedback cycles that make hard policy problems hard. Problematize need.

In a sentence, there can **never** be just one kind of sustainable development. To see whey, let us start with the Table on the following pages, which sets out the answers identified in the preceding analyses to each of the four questions.

At a first read-through, little convergence seems evident. The answers to the first and last questions—*What is sustainable development?* and *Practically, what can actually be done?*—are especially diverse, though not contradictory. What is evident is that these answers plot much of the wide range of positions on sustainable development found in the literature beyond the pages of *Ecological Applications*. The instability of desire, the need for resilience, the role of communities, the importance of justice in sustainable development—all have been touched upon at one time or another in the literature and certainly are not the special purview of the four approaches discussed here.

That said, a closer inspection of the answers indicates some initial convergence. Most clearly, answers to the third question—*Ideally, what needs to be done?*—intersect on the need to reduce the complexity and uncertainty that give rise to demands for sustainable development in the first place. (In the case of Critical theory, organic negativity is the way to get to such certainty). Simply put, we need to reduce the complexities that drive us to want to live more sustainably. Less banal, sustainable development, whatever it is precisely, is clearly a response to such complexity, not the means to reduce it. By implication, the very different answers to the third and last questions underscore the large gap between the ideal and the practical we should expect to find in sustainable development, whatever the approach, because of that complexity.

The complexity is also reflected in the various answers to *What is sustainable development?* and *Why is it an issue?* While sustainable development is many things to many people, it is complex because, wherever we stand and no matter how we look at it, it always comes back to having something to do with using resources today in ways that keep options open for their use in the future—and we really know few, if any, ways how to do that. But this is not the only reason sustainable development is complex. As the answers to *Why is it an issue?* indicate, it is complex precisely because many other issues are making it complex in addition to the drive to manage resources more sustainably for the future. Wealth, culture, managerialism, and issues of justice and much more come to complicate the matter considerably.

Table: The Four Approaches and Questions Compared

The Four Questions	Girardian Economics	Cultural Theory	Critical Theory	Local Justice Framework
What is sustainable development?	It is at best a social convention which for a time underwrites and stabilizes decisionmaking under high uncertainty in a way that its subscribers believe keeps options open for the future.	It depends on the culture. Because their views of nature and their strategies to manage their needs and resources differ so profoundly, sustainable development is necessarily defined differently by each culture.	It is a New Class version of resource managerialism that functionally serves to globalize and perpetuate the techno-managerial elite's control over everyday life. While opposed to standard economic growth prescriptions, sustainable development is really artificial negativity that leaves untouched the issue of whether or not the New Class resource managerialism is appropriate at all.	At best, it is a justice/ injustice cycle that recurrently comes back to the notion that we should manage resources today in ways that leave open future options for their management.
Why is it an issue?	Ultimately it is an issue not because people are overexploiting resources, but because some of them have reached a level of wealth where they now have the desire to stop such exploitation and achieve what they believe to be sustainable development.	Ultimately it is an issue not because people are overexploiting resources, but because people belong to cultures that are not only different, but conflicting and competing as well. What one culture takes as the answer to uncertainty--its view of nature and its corresponding management strategy--another culture takes as the source of its problems. Wherever there are multiple cultures, sustainable development will be an issue.	Ultimately it is an issue because the New Class' older versions of resource managerialism, such as economic growth and transnational exploitation, have come under increasing criticism, complexity and uncertainty, so that they now must be augmented and/or replaced by newer, aspiring versions of resource managerialism under the expert control of techno-managerial elites.	Ultimately it is an issue because local justice systems are fairly complex in practice, which means in part that the justice/ injustice cycle called sustainable development will inevitably involve elements of injustice throughout its operation

Table: The Four Approaches and Questions Compared (continued)

The Four Questions	Girardian Economics	Cultural Theory	Critical Theory	Local Justice Framework
Ideally, what needs to be done?		Ideally, the uncertainty to which culture is a response should be reduced.	Nothing "needs" to be done. Ideally, organic negativity would eliminate the "need" for New Class domination, in particular resource managerialism and its progeny, including sustainable development.	Ideally, the complexity of local justice systems should be reduced, the homogeneity of sustainable development strategies complexified, or both.
Practically, what can actually be done?	Buffer or decouple resource systems and their management from the more turbulent environments in which they are embedded. Encourage the evolution and diffusion of more than one kind of sustainable development. Differentiate sustainable development by treating it on a case-by-case basis. Resist globalizing behavior. Last, but never least, nurture both inter-local differentiation and the role of economic growth and sovereignty rights in that differentiation.	If you cannot think like a hermit, be biased in favor of your culture (and thus your own definition of sustainable development), recognizing that others are doing the same for theirs. Focus your definition of sustainable development in terms of where you stand on the usefulness of trial and error learning in development generally. Insist on the necessity of pluralism in politics; sustainable development won't be real otherwise. Be truth-seeking, whatever your culture or politics, and expect others to be the same: Sustainable development stands or falls on what is actually happening. Finally, be open to surprise and change: No sustainable development can be or, for that matter, should be once and for all or in one way only.	Protect really existing communities which seek control of their own territories. Promote a loose federal arrangement predicated on heterogeneous communities, each with the right to secede from this confederation but having no right to usurp others. Don't be deluded into thinking in terms of the Left or the Right, as they both operate according to New Class principles. Give populism a chance.	Be prepared to accommodate the reality that local justice systems are inevitable and, as such, they will *invariably* alter sustainable development strategies on the ground.

Yet when it comes to *Practically, what can be done?*, there seems to be far too many different answers for any convergence to be possible. At a very broad level, it appears that whatever sustainable development is, it should and must accommodate "difference," particularly at levels below the global or the general. It is difficult, however, to see other convergences in the answers to the fourth question from the Table only.

Fortunately, there are convergences across the preceding four chapters and analyses not reflected in the Table. The reader may wish at this stage to re-read some of the earlier material and make her or his own list of where the analyses converge. My list identifies five areas which I cast as a set of triangulated injunctions that are my own answers to *Practically, what can be done?*

Differentiate!

All four analyses insist, albeit each in its own way, that analysis must start with differentiating situations, particularly the local. Girardian economics sees local differentiation as a response to crises of undifferentiation; Critical theory insists on the pre-eminence of cultural heterogeneity of local communities; cultural theory finds an irreducible pluralism in basic cultures that define each other by being fundamentally different from each other; while the local justice framework insists that local is not global and what is local has several viable forms, be it a country, a policy arena, or a locality.

The imperative to differentiate does not, however, tell us what and how to differentiate. Other than saying the distinctions will be multiple and cross-cutting, one simply is not in a position to conclude exactly what those distinctions are. To do that, you would need a theory, not a triangulation.

That said, making distinctions, whatever they are, from the very start of one's analysis has profound implications for the sustainable development debate. First, differentiation works against the kind of globalizing discussion that pervades much of the *Ecological Applications* debate. The global level of analysis is simply not variegated enough to capture the kinds of "local" identified in the preceding chapters as crucial. (Alternatively, if one insists on keeping the discussion at the global level, then much more needs to be done to differentiate the "global" than has been the case in the *Ecological Applications* debate.) Second and more important, the imperative to differentiate implies that if you differentiate what is local from the outset (whatever your metric of differentiation), you will **never** find that "Halt population growth!" and "Reduce percapita consumption" hold true everywhere. Moreover, what you will find will be much more useful, in that differentiation means each "case" has its own merits and must be evaluated so.

This drive to analyze each case according to its merits has one very significant policy implication. Those who advocate reducing population growth often respond to those instances where increased population has actually led to *decreased* environmental degradation or increased biomass (e.g., Fairhead and Leach, 1995; Tiffen *et al*, 1994) by countering, "Surely you can't generalize from a handful of such cases!" "Yes, that is right," we ought to reply. "That is *precisely* right. We—and that includes you—can not generalize. Nor should we seek to do so from too few cases. So, then, why do you advocates of reduced population and consumption always generalize from your cases? What we are insisting is that you first differentiate cases before you do anything like generalization—which is why we are pointing to all these counter-examples you guys never talk about!!"

Beware Persecutory Language!

When the preceding is true, then blanket calls for "Halt population growth," "Reduce percapita consumption," "Any cause is a lost cause without a reduction in population," and the like are all persecutory, and importantly so. Each preceding chapter has noted this persecution, which goes well beyond the common complaints of First World racism against Third World peoples.

In making their calls for reduced growth in population and per capita consumption, the New Class damns any populism that would argue to the contrary; proponents of population and consumption control know who their real "enemies" are (i.e., each culture defines itself against the other cultures); these enemies—the breeders and overconsumers of the world—are scapegoated and to be sacrificed through the polarization of "lifeboat ethics" and "triage;" and even if exhortations, such as "Halt population growth!" and "Reduce percapita consumption!" were globally true and just, they surely lead to local injustices, given wide inter-local variation on the ground. The admonitions are persecutory precisely because their proponents on one hand say they take complexity and uncertainty seriously and on the other hand proceed to talk so certainly and simply about that which is so complex and uncertain. Such blanket calls for reduced population and consumption are offensive because they set out to efface a local differentiation, whose main implication is that the proffered advice cannot be everywhere the case.

Ironically, one answer to the population and consumption problems identified by sustainable development advocates is for these advocates to stop hounding people who don't obey their warnings. That would not solve the "problem" they see, but it would recognize that these advocates have absolutely no standing in the crisis upon which they are sermonizing. No standing? They have no standing precisely because their statements of Halt population growth! and Reduce per capita consumption! have zero—repeat, *zero*—policy relevance in the absence of these advocates providing estimates on what levels of population and consumption have to be in place so that people can have markets, participate in their development, and manage their own resources sustainably. The advocates simply do not know enough to provide these estimates on the variable and differentiated scales required—and *they never will.*

Acknowledge the Unmanageable!

Each of the preceding chapter analyses recognizes that an important part of what managers want to manage—or believe they must manage—is itself unmanageable and uncontrollable. This acknowledgment is especially meaningful for a sustainable development that places management at the core of its definition and practice. In Critical theory, we saw the unmanageable in organic negativity. In Girardian economics, it took the form of a rapidly cascading crisis of undifferentiation. In the local justice framework, we saw the impossibility of designing a compensatory management mechanism to rectify the global injustices accumulated through the workings of locally just systems. In cultural theory, we wondered at the difficulty of designing and managing pluralist policies around cultures that see themselves as mutually exclusive and centrifugal in all the important respects. In each case, there is a point at which managers find they cannot manage what they want (have) to manage.

The more managers (want to) manage, the more they (have to) confront the unmanageable. But it is equally true that the more unmanageable things are, the greater the pressure to manage them. "Few managers," as Demchak (1991, 31) puts it, "will let a disturbance with seriously negative consequences simply exhaust itself." When confronted with the uncontrollable, these managers "exhibit a bias toward greater control" (Op Cit). This pressure to control is certainly at the heart of a sustainable development, which is nothing if not the antithesis of a wait-and-see attitude that assumes something will always come along to deal with our problems, or these problems can't matter that much. Sustainable development in this way ends up embracing a management feedback cycle that undermines its potential for achievement—management ultimately confronts that which cannot be managed; but that which cannot be managed leads to greater pressures to manage. Or for our purposes, efforts to manage sustainably quickly encounter a complexity that is in important respects unmanageable; but a complexity that has not been (cannot be) managed puts even more strain on managers to try to control it—and as soon as possible.

Search for Feedback Cycles!

This management/unmanageability feedback cycle is not the only one that matters for sustainable development. The tissue and muscle of sustainable development are hard policy issues, and these issues are hard not merely because they are complex and uncontrollable, but because they also involve a whole range of other feedback cycles that are not easily broken into and correctable.[6] Each of the four preceding analyses identified its own "perverse" feedback cycle—differentiation/undifferentiation, overregulation/underregulation, opposing cultures/inevitable pluralism, local/global justice and injustice. Many more exist (see, e.g., Hukkinen, Roe and Rochlin 1989). The more you search complexity, the more perverse feedback cycles you will find. And the more feedback cycles you find, the more you will understand why sustainable development is *necessarily* so difficult. In fact, feedback cycles of complex

---

[6] According to the Ehrlich and Daily response (559) in *Ecological Applications*, "our interacting ecology and economics group at Stanford has discovered that it comes to rather easy mutual agreement" on a number of hard issues. Something must be wrong at Stanford.

interactions that are coupled tightly (Cell 2), loosely (Cell 4) or a mix of both are prevalent throughout policy life and are the defining moment of our understanding of why these issues are complex and difficult at the same time.

The triangulation reminds us that, no matter where you stand or how you look at it, sustainable development embraces hard issues that are going to be around for some time, *especially* as the world gets increasingly complex.

Problematize Need!

As noted in an earlier chapter, the rhetoric of needs is difficult to avoid and the foregoing pages prove that true. Still, these analyses have little patience with action programs and their analytic frameworks that are based on underspecified, abbreviated understandings of what people "need." In none of the preceding analyses is "need" ever straightforward. For Girard, markets create needs; for Piccone, so many of our needs today are ones of artificial negativity cultivated by our New Class gardeners; for Elster, level and increment needs are often insufficiently specified; and for Douglas and Wildavsky, need is fundamentally dependent on one's way of life, and ways of life vary always. Whatever the merits of these individual arguments, their convergence underscores that "human needs" are rarely as simple as those discussed in the pages of *Ecological Applications* and *Science* or by many sustainable development advocates. Fortunately, as we saw in Part I, abilities matter more than needs in a complex world.

***

Other convergences may well exist and I would be interested in the reader's list. The above five are enough, however, to establish one fairly important feature of triangulation as a method. Some readers might be thinking that the five injunctions are merely an artifact of the four approaches used. Other theories would have produced other lists. Possibly. But never entirely, at least when triangulation is as successful as this one.

To see why, step back a moment and look at the keywords in the above injunctions—"differentiate," "feedback," "problematize," and "unmanageable." They in fact are synonyms and cognates for the keywords in our definition of complexity. Complexity, to repeat, has three elements—the number of components in a system, their degree of differentiation, and their degree of interdependence—and one primary outcome, surprise. These features are precisely what is being converged on in the triangulation, though in a much more relevant way than the conversation stopper, "complexity is complex," indicates. If you will, policy relevance is giving a face to *in situ* complexity which, in the absence of triangulation, has no discernible face.[7] Or to return to

---

[7] In other words, although we would expect overlap, the same approaches applied to an entirely different issue might well not produce exactly the same set of answers as "differentiate!," "problematize needs," "acknowledge the unmanageable," and "search for feedback cycles." It all depends on the complexities being analyzed and what part of the approaches are accented in the analysis. For a triangulation using the four approaches but

our earlier metaphor, we started in complexity, and we have ended in complexity, the only difference being that between start and finish we used a compass to chart out a terrain we feel more confident in crossing.

With this, we are now in a position to describe the nature of the "convergence" in triangulation more clearly. What we are converging on is not "a common ground" between the approaches. After all, they have been chosen precisely because each opposes the other in fundamental respects. Rather the triangulated answers to the four questions that have driven Part II are better thought of as describing complexity in a way that we can now do something about that complexity.[8] More formally, the statement, "The four approaches are incommensurable (roughly orthogonal) and therefore no common ground is possible," assumes a metalanguage in which we can compare and contrast these approaches and come to the conclusion both that the statement in quotes is objectively the case and *just how it is the case.* This is a metalanguage of comparison and contrast that takes complexity as complexity (i.e., describes pluralism as more than the sum of its worldviews), and, as discussed in Part I, triangulation is a central component of that metalanguage in practice.

CONCLUDING REMARK

The triangulated answers reconceive the *Ecological Applications* debate and the wider sustainable development in a profound way. To see how, start where the *Ecological Applications* debate all began, the first sentences of the Ludwig *et al* article:

> *There are currently many plans for sustainable use or sustainable development that are founded upon scientific information and consensus. Such ideas reflect ignorance of the history of resource exploitation and misunderstanding of the possibility of achieving scientific consensus concerning resources and the environment. Although there is considerable variation in detail, there is remarkable consistency in the history of resource exploitation: resources are inevitably overexploited, often to the point of collapse or extinction. We suggest that such consistency is due to the following common features: (i) Wealth or the prospect of wealth generates political and social power that is used to promote unlimited exploitation of resources. (ii) Scientific understanding and consensus is hampered by the lack of controls and replicates, so that each new problem involves learning about a new system. (iii) The complexity of the underlying biological and physical systems precludes a reductionist approach to management. Optimum levels of exploiting must be determined by*

---

in a slightly different way and on an entirely different case, see Appendix 2.

[8] In narrative policy analytical terms, the metanarrative tells another story altogether than the opposing narratives triangulating on it.

trial and error. (iv) Large levels of natural variability mask the effects of overexploitation. Initial overexploitation is not detectable until it is severe and often irreversible.

In such circumstances, assigning causes to past events is problematic, future events cannot be predicted, and even well-meaning attempts to exploit responsibly may lead to disastrous consequences...

What's wrong with this statement? From the view of the critics in *Ecological Applications*, Ludwig *et al* are right about uncertainty and complexity, but wrong about the rest. For all of those involved, this is a debate over issues of fact and informed judgment, and they cite what they take to be counter-examples to the Ludwig *et al* position.

But that is not what's wrong with the statement from the view of the preceding four approaches. They too start in the midst of uncertainty, but come to very different conclusions about what are the relevant issues for sustainable development. *Nor by this point should these other views be any more remote or implausible than the critics' response.* Once you know the four approaches, you see each reflected in the above statement. Girardians immediately pick up the phrases about "wealth," "irreversible," and "overexploited...to the point of collapse or extinction," seeing in them the slippery slope of mimetic rivalry and crises of undifferentiation. Critical theorists for their part see all manner of New Class aspirations in the "many plans for sustainable use or sustainable development". Cultural theorists see the play of very different cultures in that fatalistic "resources are inevitably overexploited" versus the hierarchist belief that "optimum levels of exploitation must be determined by trial and error." As for advocates of the local justice framework, the possibility of local justice systems leading to global injustices is glimpsed in clauses such as "even well-meaning attempts to exploit responsibly may lead to disastrous consequences".

If we were to leave it at that, the conclusion would be that any number of different ways are possible to look at Ludwig *et al*'s statement, and by implication the sustainable development debate generally. At one level, that is true. But triangulation allows us to by-pass that conclusion when the approaches actually converge. From the perspective of our triangulation (and thus from the perspective of the four approaches considered together), what is really wrong with the Ludwig *et al* statement can be discerned *not in the above excerpts but in an entirely different point*, namely their conclusion that "complexity of the underlying biological and physical systems precludes a reductionist approach to management."

From the perspective of the triangulation, is it not ironic that ecologists, who take such care in arguing against reductionism in the face of biological and physical complexity, plow ahead and argue for just such a reductionism in the face of social and economic complexity? For what else are statements such as "halt population growth" and "reduce percapita consumption," if not single-minded reductionism in the social and demographic realm? Indeed, what is more reductionistic than sustainable development advocates who see in sustainable development an end to envy, an end to crises of undifferentiation, an end to the conflict that comes with cultural pluralism, an end of negativity, and an end to discretion in local decisionmaking? Ludwig *et al*—as indeed many of their critics—are quite content indulging in big-picture simplifications when insisting on talking globally and generally (thereby ignoring local differences and specific case-by-case analysis), treating human needs and behavior as unproblematically straightforward (e.g., "wealth...generates... power"), and thinking the best way to adapt to the unmanageable is to gain control of it through trial and error. Yet none of us should be surprised that the socio-demographic world is considerably more complex than that—perhaps *even more complex* than biological and physical systems, if we take the earlier definition of complexity as the metric.[9]

So then are the conclusions of the triangulation. But what do these conclusions leave us for practicing sustainable development on the ground? How do these insights, including those in Part I, change our lives and our practices with respect to keeping resource options open in the future? Part III provides the answer.

---

[9] As we saw earlier, reductionistic approaches per se are not ill-founded; in fact, they are needed in triangulations. The problem is when such approaches become the only approach when analyzing issues of high complexity.

APPENDIX 1: MODELS OF INTERDISCIPLINARITY

The more complex have become policy problems and policymaking related to natural resources and the environment, the more interdisciplinary the approaches and modeling recommended for improving that policymaking. The *Ecological Applications* respondents call for an interdisciplinary focus on sustainable development for just this reason—namely, sustainable development is much too complex for just one discipline to handle alone (e.g., Holling 1993; Ehrlich and Daily 1993).

Ten different approaches to interdisciplinarity command attention today (see Roe 1994c):

1) The *Synergy Model*. Here the analogy is the heterosis effect, i.e., in the process of working together, officials, professionals and academics from different disciplines produce something new or innovative. Example: the experience of some team-taught workshops and courses.

2) The *Puzzle Model*. The analogy is a 500-piece picture puzzle, with one profession bringing x pieces to the board, another y pieces, a third z pieces, and so on. Example: many reports of multidisciplinary task forces.

3) The *Triangulation Model*. The analogy is that of multiple searchlights intersecting in the night's sky. Example: use of very different theories to throw more light on some aspect of an otherwise murky issue, without expecting the whole picture to become clearer (as in the Puzzle Model) or that something new will emerge out of the intersection (as in the Synergy Model).

4) The *Cross Application Model*. The analogy is that of importing a framework or model developed in one discipline for use in another discipline where it has not hitherto been applied. Example: a great deal of applied social science theory for ecosystem management discussed in Part III.

5) The *Interdisciplinarian Model*. Here it is the analyst—official, professional, academic, other—who is trained in several disciplines. Example: students in graduate schools of policy analysis taught a common analytical toolkit, i.e., microeconomics, research methods, organization theory, legal analysis, and public management practice.

6) The *Key Informant Model*. Rather than the analyst being trained in several disciplines, it is those being analyzed who are interdisciplinary. Example: interviewing key informants or those observers having "the big picture."

7) The *No-Disciplines/Only-Interdisciplinarity Model*. Here the analogy is the withering away of conventional disciplines or professions into those said to be inherently interdisciplinary. Example: environmental studies, irrigation management.

8) The *Ad Hoc Model*. Here the assumption is that any arbitrarily assembled group of dedicated professionals has a chance of coming up some effective ways of approaching the policy problem of interest, albeit a different group altogether would come up with different answers that could be just as effective. Example: advertising campaigns.

9) The *Coordinated Disciplines Model*. Here the disciplines and professions remain the major mechanisms for analysis, though individual practitioners work together to coordinate their various aspects of their projects. Example: researchers agreeing on a common level or scale of analysis to be used by them in their own individual work.

10) The *Serial or Iterative Model*. Any of the above models in any sequence.

Not all approaches are equally prevalent or recommended. I have worked in a college, where many faculty: (1) believe they are practicing the puzzle or Ad Hoc Models in their public service; (2) would be open to the Synergy Model in their teaching, are committed by virtue of the college's mission statement to training students in the Interdisciplinarian Model, would rather not bother with the Coordinated Disciplines or Cross Application Models, and loathe the idea of a No-Disciplines/Only-Interdisciplinarity; while (3) in their research reject these seven models altogether, i.e., they are committed disciplinarians, particularly the natural scientists. Plus the faculty don't have time for the Serial Model, albeit privately many of them see themselves as wonderful examples of the Key Informant Model.

What, though, about the *Triangulation Model*? Triangulation, as we have seen, holds out the great promise for analyzing precisely those kinds of complex policy issues such as sustainable development. Triangulation illustrates better than any of the other models the strengths and limitations of policy-relevant interdisciplinarity. Interdisciplinarity's chief strength is that it is a useful way to start out the analysis of difficult policy issues. In particular, triangulation is both the search for a convergence on what to do generally and at the same time the recognition that much more will have to be understood specifically, case-by-case.

The chief limitation of all interdisciplinary models (including triangulation) is that interdisciplinarity is easily confused with analyzing each case on its own merits. It may be the way we begin the analysis of each case, but interdisciplinarity tells us nothing about how to complete case-by-case analysis. Composite theories may be wonderful examples of analyzing each case in its own right, but sadly they are poor templates for analyzing future cases in their own right. Different situations will require different composite theories and, as such, no composite theory can be general enough to determine where and when other composite theories should be applied.

That said, the great virtue of triangulation over the other models is its up-front insistence that policymakers will always be judged in terms of the management problems they try to solve and the solutions they enact, not whether they were "interdisciplinary" per se throughout. Their work as policymakers stands or falls not on whether they are interdisciplinary, but rather on whether they can better analyze issues of high complexity and uncertainty. Unlike the other interdisciplinary models, which treat interdisciplinarity as an end, triangulation is only a beginning, but a good beginning if the real end is analyzing problems that are complex all the way down.

APPENDIX 2: INCIDENT AT MUNDUR (EXCERPTED AND EDITED FROM DAMODARAN AND ROE, 1997)

Background

Mundur, a semi-arid Indian village within the Karnataka district of Bangalore, has been a victim of periodic droughts throughout its history. Villagers, who practice a rainfed agriculture with little opportunity for irrigation, typically lose much of their crops and many of their livestock in the droughts. In more recent times, the village has had the occasional benefit of both state and national government programs, particularly government-sponsored drought and work relief schemes. These programs have been the fairly standard "one size fits all," and Mundur has probably not benefited any more or less from them than other villages with the same droughts and programs.

Villagers also have "standard" responses to drought. This Appendix focuses on one response, in this case a ritual to appease Varuna and the weather deity, Parjanya, so that they would bless the village with rains. One such ritual was held in August 1976 and another in July 1984. Available information suggests that both incidents did not substantially differ in term of how they were enacted. There was one major difference, however, that being the community action which followed the 1984 ritual. More about that momentarily. First, the ritual needs to be described.

The 1984 ritual took place after the failure of May rains considered crucial to the success of land preparations and sowing of the staple crop, finger millet. The monsoon had failed, the villagers were demoralized, and demands for government assistance were increasingly strident. Government help to farmers in drilling bore-wells was especially sought after, as bore-wells were considered the only technologically reliable, year-round water source, in this case for irrigated rather than rainfed agriculture.

Such calls for help should have been unnecessary, as government policy was to assist farmers in meeting their minor irrigation demands. Not so for Mundur and adjacent villages, as they fell within an area designated as suffering an overdraft of groundwater and thus not covered by financial assistance programs that were seen to exacerbate the overdraft. The only exemptions to the embargo on government-assisted bore-well drilling were for the lower castes and drinking water purposes. The lower castes typically had insufficient holdings to justify assistance and Mundur did not have a drinking water problem, even during the drought. The pressing problem was the shortage of irrigated water for holdings of sufficient size to use that water. The demand for bore-well irrigation water was thus greatest among landlords and large landholders.

A week before the ritual a group of villagers, including three who were considered the core village leadership, the temple priest (who was also an agriculturalist), and a leader of one of the lower castes came to the Village Accountant's Office, where one of us (Damodaran) was also present as a central government trainee. The villagers informed them about their intention to hold a ritual to propitiate the gods, Varuna and Parjanya. A date had been chosen that was deemed astrologically auspicious. Not only were contributions requested from the village accountant and central government trainee, but their participation was sought as well in both the ritual and other connected activities left unspecified at that time.

Two things were evident early on. The three village leaders where the principal financiers of the ritual and the ritual would be open to everyone in the village without reference to their caste, creed or class. Invitations had gone out to relatives from two of the nearby villages to participate as well.

Early in the morning on the day of the ritual, the village drummer beat his drum inviting villagers to congregate at the Saivate Temple (dedicated to the god Shiva). By an hour later, a huge congregation had assembled at the temple, including the village's core leadership, the lower caste leader, the village accountant and central government trainee. The priest was inside the temple's inner sanctum for another hour offering prayers. Meanwhile outside the temple two donkeys were brought forth with bright red velvet on their backs. Two sacks of frogs were also kept to the side.

After his prayers, the priest came out and offered the camphor flame for those who had thronged to the temple. One of the village's core leaders applied sandal paste and red powder to the donkeys' foreheads, while another of the leaders poured water into the sacks of frogs. The lower caste leader placed the wet sacks on the donkeys' backs and then led them and a procession of enthusiastic and excited villagers dancing, drumming and chanting around the temple three times.

The procession then moved to a site away from the temple where a dais had been improvised for speakers. One core leader started the speeches by reviewing the conditions that gave rise to the ritual. In contrast to the 1976 ritual, he went on to encourage his audience to march the next day to the Block Development Officer (BDO), who was responsible for sanctioning new bore-well construction in the area. The purpose of the march would be to encourage the BDO to approve a new drinking water bore-well for Mundur.

At this point, the village accountant raised objections and cautioned against the proposed march, suggesting instead that the villagers make representations to the BDO only when he next came to the village. The central government trainee, who was not under the local or state government chain of command, countered that there was no harm in presenting the village's grievances in the form of such a march. The march did take place the next day. The BDO assured the marchers that he would consider their request, although in the end he did not approve it. Instead he sanctioned the desilting of the village ponds, an intervention that not only employed people but could be justified on grounds of the overdraft problem.

It turned out the reasons for the march, and by implication the preceding day's ritual, were murkier than initially supposed. Apparently another factor also explained why the core village leadership had wanted the new drinking water bore-well. All three had tried drilling their own private bore-wells, succeeding however in only four out of 10 attempts. Had the new bore-well been drilled successfully, the villager leaders reasoned that this would be an indication that groundwater reserves were in fact sufficient to merit even more bore-well drilling, in this case for private purposes.

While the 1976 and 1984 rituals were enacted on the same lines, what followed the 1976 ritual was not a march but a massive communal effort using unpaid village labor to desilt the village pond. In contrast, the desilting of the village ponds that followed the 1984 ritual was undertaken by villagers who were paid as part of the government assistance program.

Those are the details. The following triangulation is aimed at explaining why this ritual and subsequent march took place, given the uncertainty over what were the real reasons for both. Was it primarily because of the drought? Did the material interests of the core village leadership drive both events into happening or rather did these interests merely help the events along? Was it also important that one of us was actually there, contingently in the right place at the right time, to recommend that the march to the BDO actually take place? What else mattered? Answers to such questions are important because of their major policy implications. If we know why people follow up the same action (a ritual) with different behavior (in 1984 a march, in 1976 not so) under very similar natural disasters (both severe drought years), we will have a better understanding of what can and cannot be done, politically and economically, in response to such disasters.

Application of the Four Approaches

Each approach is applied to the incident at Mundur and the triangulation follows those applications. The following differs in two important respects from this chapter's preceding exegesis and triangulation. Because the focus here is on explaining a ritual, we have relied directly on the work of René Girard, who has spent considerable time explaining the role of mimetic rivalry in the origins of ritual behavior. We have as well relied on Gross and Rayner's exposition of cultural theory, for they specifically discuss that theory in terms of developing country examples.

Girardian Analysis of Ritual Behavior and its Origins in Persecution

First, René Girard's theory:[10]

> No matter what circumstances trigger great collective persecutions, the experience of those who live through them is the same. The strongest impression is without question an extreme loss of social order evidenced by the disappearance of the rules and "differences" that define cultural divisions....We can then speak of a stereotype of crisis which is to be recognized, logically

---

[10] Girard 1989 (1986), 12, 14; 14, 21; 17, 18, 21; and 24.

and chronologically, as the first stereotype of persecution. Culture is somehow eclipsed as it becomes less differentiated.

[The second stereotype] Men feel powerless when confronted with the eclipse of culture; they are disconcerted by the immensity of the disaster but never look into the natural causes...But, rather than blame themselves, people inevitably blame either society as a whole, which costs them nothing, or other people who seem particularly harmful for easily identifiable reasons. The suspects are accused of a particular category of crimes....In order to blame victims for the loss of distinctions resulting from the crisis, they are accused of crimes that eliminate distinctions.

I turn now to the third stereotype. The crowd's choice of victims may be totally random; but it is not necessarily so. It is even possible that the crimes of which they are accused are real, but that sometimes the persecutors choose their victims because they belong to a class that is particularly susceptible to persecution rather than because of the crimes they have committed... There are very few societies that do not subject their minorities, all the poorly integrated or merely distinct groups, to certain forms of discrimination and even persecution....[I]n actuality they are identified as victims for persecution because they bear the signs of victims.

Each time an oral or written testament mentions an act of violence that is directly or indirectly collective we question whether it includes the description of a social and cultural crisis, that is, a generalized loss of differences (the first stereotype), crimes that "eliminate differences" (the second stereotype), and whether the identified authors of these crimes possess the marks that suggest a victim...(the third stereotype). The fourth stereotype is violence itself....The juxtaposition of more than one stereotype within a single document indicates persecution. Not all stereotypes must be present: three are enough and often even two. Their existence convinces us that (1) the acts of violence are real; (2) the crisis is real; (3) the victims are chosen not for the crimes they are accused of but for the victim's signs that they bear, for everything that suggests their guilty relationship with the crisis; and (4) the import of the operation is to lay the responsibility for the crisis on the victims or at least by banishing them from the community they "pollute".

For Girardians, all rituals originate in such persecution of victims (see, e.g., Girard 1987, 178). Over time, those involved in rituals lose sight of the origins in sacrifice and violence against scapegoats, but each time the ritual is enacted its objective is to banish whatever is perceived to being polluting the community in a way that is also perceived to end the crisis and restore order: "The community must effectively be emptied of its poisons. It must feel liberated and reconciled within itself. This is implied in the conclusion of most myths. We see the actual return to the order that was promised by the crisis, or even more often the birth of a new order in the religious union of a community brought to life by its experience" (Girard 1989 [1986], 42). "Admittedly, scapegoats cure neither real epidemics nor droughts nor floods;" Girard goes on to write (Ibid, 43):

> As long as external causes exist, such as an epidemic of plague, for example, scapegoats will have no efficacy. On the other hand, when these causes no longer exist, the first scapegoat to appear will bring an end of the crisis by eliminating...all evildoing in the person of one victim. The scapegoat is only effective when human relations have broken down in crisis, but he gives the impression of effecting external causes as well, such as plagues, *droughts*, and other calamities. [our italics]

From a Girardian perspective, the incident at Mundur is just such a ritual. Who the original scapegoat was and what form that sacrificial violence took need not detain us. Suffice it to say, the village was the victim and the crisis was a drought that left a once a varied rainfed agriculture now everywhere undifferentiated. As pouring water into the gunny-sacks of frogs, the ritual sought to restore order by expunging the drought that had blighted the community. In the "Rig" Vedic tradition, the croaking frogs symbolize the onset of rains (i.e., the end of the drought), while the donkeys symbolically carry the burden of atoning for villager acts of omission and commission in the past. The ritual's bloody origins were visible only in the simulacrum of mob-like behavior of the temple procession along with the red of the velvet and powder.

Not surprising to Girardians, the ritual led to something novel, in this case the march to the BDO's office, as a crisis of undifferentiation leads to rise of new social conventions. Moreover, for Girardians the function of any new convention arising out of the crisis is to re-differentiate a reality left undifferentiated by that crisis, and indeed the march was to secure a different type of water supply for a different type of agriculture than had been common in the past.

The Local Justice Framework

What went on at Mundur was an example of a local justice system at work. Villagers in parts of arid and semi-arid regions of India exercised a fair degree of autonomy both as a community and as operators in an agricultural sector that can be weakly controlled by state or central governments. Moreover, the distribution of drought-related deterioration is non-monetary in the direct and formal sense. As agriculture is basically subsistence oriented (though some crops are sold for commercial reasons), the drought-induced decline in yields is not immediately in monetary terms. The same too holds true for grazing. As livestock are typically held not for cash nor for meat but for other reasons (draft, manure, and milk), the effect of the drought on grazingland is also not immediately in monetary terms. Local justice is moreover only partially compensatory here. Larger livestock holders do not, as a rule, directly compensate smallholders for the shortage of grazing on the commons caused by their larger herds. Nor do livestock holders compensate non-holders who cannot start or maintain herds because of insufficient grazing or water (both surface and ground) left for them to do so.

Local justice systems are not, to remind the reader, designed to correct for global injustice, which is certainly true for agricultural systems, and not just in semi-arid India. Agriculture systems do not, as a rule, compensate a villager who has had a string of bad luck, e.g., born poor, left ill-educated, allocated little land, never holding more than one or two livestock and always losing those in the next drought. Yet the financial assistance program for bore-wells was designed to rectify just such a global injustice, in this case, the "bad luck" of being born a lower caste or without drinking water in a semi-arid environment that is vulnerable to recurrent and devastating drought. The injustice here is global precisely because whatever the agro-pastoral system—communal, private, leasehold or mixed—it is not one capable of compensating for this kind of "bad luck" on its own. The march to the BDO was, in effect, an effort of one local justice system to seek administrative redress (i.e., a bore-well) for its own agricultural limitations through recourse to a global justice system whose programs, in this case, included cash and financial transfers as compensation (e.g., through work-relief programs).

Cultural Theory

Cultural theory implies that the "rational peasant" is never rational or homogeneous with all other peasants in the same way. Villagers must be differentiated even within the same village, and in ways that go well beyond the wealth and caste categories of rich and poor, high and low. The various "actors" in the Mundur can be located in a grid/group typology as follows:

Grid

	Group	
	Low	**High**
High	D Village Accountant & BDO (Fatalist)	B Villagers (Hierarchist)
Low	Core leaders A (Individualist)	C Government trainee (Egalitarian)

The low grid/low group cell, in the words of Gross and Rayner (1985, 6), "allows the maximum options for negotiating contracts or choosing allies. Consequently, it allows for individual mobility up and down the scale of prestige and influence. In this environment, a frontiersman or a capitalist entrepreneur might thrive..." The three core village leaders come closest to meeting this description in their drive to secure private bore-wells and are, accordingly, placed in the individualist quadrant.

The popular view of traditional village society fits the high grid/high group cell, where, again in the words of Gross and Rayner (9), "one might find tradition-bound institutions in which everyone knows his place...Examples of this type of social organization include...a cohesive tribal society with hereditary roles." Consequently, villagers have been assigned to that quadrant. (The priest would normally fall within the hierarchist quadrant, though his position as an agriculturalist and subsequent material interest in the bore-wells might more properly put in him the individualist cell. As we will see in a moment, there is a level of analysis at which it does not matter to which of these two quadrants the priest is assigned.)

The low grid/high group cell "represents a social context in which the external group boundary is typically the dominant consideration. All other aspects of interpersonal relations are ambiguous and open to negotiation. Leadership tends to be charismatic and lacking clear rules" (Gross and Rayner 1985, 10). The central government trainee comes closest to meeting this description, as he supported and encouraged the villagers in mobilizing their march, but did so as one less subject to the same prescriptions and conventions under which the villagers and local government operated. Finally, while none of the principals to this specific event behaved fatalistically, at least in the conventional sense of the word, the relatively individualistic behavior of the village accountant and BDO, both of whom however operated under a greater concern for government protocol and regulations, puts them closest to the high grid/low group cell.

Ironically, this fatalist quadrant allows us to test the efficacy of the cultural theory framework in explaining the incident at Mundur. According to Gross and Rayner (9), high grid/low group examples include some of the lower castes in India. It could be argued then that Mundur's lower castes should be so classified as well. Yet, during the ritual and subsequent march, they acted in a high grid/high group fashion, as described earlier. Thompson, Ellis and Wildavsky (1990, 76) provide examples of when individuals shift from fatalist to hierarchist:

> From fatalist to hierarchist is Marx's "dictatorship of the proletariat." His description, of course, assumes this to be a macrochange (one that is taken by all people as they come together to cast off their chains)...

This description, with only a few changes, neatly fits what happened during the Mundur incident.

As in our earlier example, quadrants along one diagonal share something in common that those along the other do not. The core leaders and villagers along diagonal A/B are the *insiders*, while the village accountant and BDO along with the central government trainee along diagonal C/D are the *outsiders*, at least with respect to Mundur's immediate water problems. The Village Accountant, for example, was not from the village at all. The village insiders took up and participated in the ritual and its aftermath in a way the village outsiders did not. In this sense, for example, it does not matter where the village priest is placed along diagonal A/B, as he acted throughout as a village insider.

This insider/outsider dimension to the incident helps explain one seeming anomaly in what happened. If an aim of the core leadership was to secure new and hopefully positive drilling results as a way of enhancing their own chances for later obtaining private bore-wells, why did the leaders not help to ensure that one or two of the lower caste villagers could qualify for bore-well assistance to which they were specially entitled? Such lower-caste support would seem to have been all the more appropriate, as it was generally recognized that the village could not justify new drilling on the grounds of a drinking water need only. One answer is to appeal to the differing material interests of the castes involved. Another answer is appeal to cultural differences, e.g., lower castes are more "fatalistic," though as we just saw that explanation has problems for the event under question. A more parsimonious answer is to assert that drilling for drinking water purposes was the one thing that village insiders could agree on in the face of a bore-well assistance program defined and controlled solely by outsiders, such as the BDO.

Critical Theory
From the perspective of Critical theory, the incident at Mundur can be explained as the techno-managerial elite coopting an organic (i.e., populist) ritual for the instrumental ends of the New Class bureaucrats concerned. The ritual was originally organic negativity (a.k.a. populism) at work, in this case, villagers coming together as a group on their own to address a common crisis without recourse to a government—bureaucratic—sanction that there was a crisis and that it should be responded to in a certain way. Members of the New Class in this specific incident—the village accountant, the government trainee and the BDO—may have disagreed among themselves, but whatever their intentions the functional effect of their actions was to reinforced their elite status as the "experts" who knew how to "solve" the villagers' water problems through their—the New Class'—better advice, management and/or technology.

Consequently, the march on the BDO's office was functionally an example of artificial negativity coopting the organic negativity of village populism. At first glance, the march may have looked like a villagewide "counter-challenge" to the government assistance program, but in reality the march served to reinforce the perceived need for such a program and legitimize the role of the BDO in managing such programs. Marches are thus rendered into just one more reason BDOs and their state efforts are required and, in turn, seek to justify themselves. That the march was coupled with a ritual involving the whole ("organic") community served only to enhance the legitimacy of the villagers' request, which in turn only enhanced the legitimacy of the New Class programs and bureaucrats who could choose to meet that request and in what manner. Indeed, the fact that the 1984 ritual led to a government program paying villagers to desilt their ponds, whereas in 1976 the desilting was done by the unpaid villagers communally and outside any government program, indicates just how far and deep had penetrated the New Class ideology of a one-size-fits-all package of government entitlements and services that serialize villagers into abstract individuals shorn of their populist and particularistic character.

The Triangulation

Fortunately in the case of the Mundur incident, the four approaches triangulate on a shared point of departure for further follow-up, which can be stated without much ado: What connects the four approaches *to this case* is the need from the very beginning to differentiate the participants in this event and the context(s) in which they operate, if the aim is policy relevance.

Whatever the theory used, the implication to be drawn is that these people and the situations in which they find themselves must be distinguished from each other, and from the outset of any such analysis. The point is not that these distinctions are specifically local/global, organic/artificial, differentiated/undifferentiated, individualist/others (indeed, other theories would offer entirely different distinctions), but that such distinctions are multiple and cross-cutting in the political and economic situations that most matter to the people concerned.

Thus, from the perspective of the four approaches used in *this* triangulation, the central government trainee was not just an outsider to the village, but in the same instant he was also a member of the techno-managerial elite, an egalitarian, and a principal participant in the generation of a new social convention that arose out of the ritual. These distinctions are cross-cutting in the sense that a participant to the event may share some, but not all, of the characteristics of other participants. Cross-cutting distinctions could as well be generated for the other participants in the Mundur event (e.g., the three core village leaders were village insiders, who were as well important individualists but who were not among the techno-managerial elite, etc.). Again, the point is not that these specific cross-cutting distinctions are what mattered the most for understanding the incident, but that multiple and cross-cutting distinctions have to be identified from the outset when analyzing situations that matter most to the people being analyzed.

The empirical finding that villagers and those with whom they interact (in the Mundur case, government officials) are frequently differentiated is, of course, not new. We saw it earlier in the admonition, *Differentiate!* What is new is the methodological imperative that insists the analyst first differentiate these participants—and differentiate them in multiple and consequently cross-cutting ways theoretically—before evaluating what the "crisis" is to which they seem to be responding. To put it more forcefully: If you differentiate these people from the beginning, what you will find is not just "a response to drought," but also other "crises" at work, including in our case, co-optation by the techno-managerial elite, insiders versus outsiders, and a clash of different cultures, to name but three.

What *this* triangulation converges on in the recognition of multiple and cross-cutting distinctions is the admission that not only is each policy-relevant situation complex, but it may well also be its own case. When this is so, the major policy implication has to be sobering: The more complex politics and economics seem at the microlevel, the more governments respond by seeking standardized approaches to the concomitantly increasing complexity at the macrolevel (Roe, 1994b). Whether rightly or wrongly, when policymakers increasingly find local complexity everywhere, they respond by insisting that the now widespread complexity requires a widespread approach to its management or amelioration. Yet no government program could ever hope to adequately deal with all local (in reality, localized) complexity. Thus, the central paradox of all development strategies: The more each case should be judged on its own merits because of its complexity, the more government responds by insisting that one-size-fits-all programs are required for that complexity. Once again, as we saw in Part I, a Cell 2 and 4 world calls forth Cell 1 thinking.

In sum, government programs, such as those for drought assistance, will *inevitably* fall short of local requirements. These programs must fall short, as local people bend, twist, resist and otherwise subvert them to meet local contingencies. In such a world, the best one can hope for is that one-size-fits-all programs learn from this experience. What damns such programs is not that they fail to implement what they plan (this failure, again, is inevitable); it is when they do not learn from such failure.

PART II REFERENCES

Akerlof, G. 1984. A Theory of Social Custom, of Which Unemployment May Be One Consequence. *In* An Economic Theorist's Book of Tales. Cambridge University Press, Cambridge.

Banerjee, A. 1992. A Simple Model of Herd Behavior. Quarterly Journal of Economics **107**: 797-817.

_____. 1993. The Economics of Rumours. Review of Economic Studies **60**: 309-327

Bikhchandani, S., D. Hirshleifer and I. Welch. 1992. A Theory of Fads, Fashion, Custom, and Cultural Change as Informational Cascades. Journal of Political Economy **100**: 992-1026.

Brewer, J. and A. Hunter. 1989. Multimethod Research. Sage Library of Social Research 175. Sage Publications, Newbury Park, CA.

California Urban Water Agencies. 1995. Ecosystem Management: The Bay-Delta Solution. Final Report (July 1995). Sacramento, CA.

Chambers, R. 1992. Rural Appraisal: Rapid, Relaxed and Participatory. IDS Discussion Paper. Sussex, England.

Conlisk, J. 1980. Costly Optimizers versus Cheap Imitators. Journal of Economic Behavior and Organization **1**: 275-293.

Cook, T. D. 1985. Postpositivist Critical Multiplism. *In* Social Science and Social Policy (R. Lance Shotland and Melvin Mark, Eds.). Sage Publications, Beverly Hills, CA.

Costanza, R. 1993. Developing Ecological Research That is Relevant for Achieving Sustainability. Ecological Applications **3**: 579-581.

D'Amico, R. and P. Piccone. 1992. Introduction. Telos 91 (Spring).

Damodaran, A. and E. Roe. 1997 Theorising to Explain and Triangulating to Explain Away? The Art and Non-Art of Multimethod Policy. Economic and Political Weekly.

Demchak, C. 1991. Military Organizations, Complex Machines. Cornell University Press, Ithaca.

Denzin, N. 1970. The Research Act. Aldine Publishing Company, Chicago, IL.

Doherty, J., E. Graham, and M. Malek. 1992. Postmodernism and the Social Sciences. St. Martin's Press, NY.

Douglas, M. and A. Wildavsky. 1982. Risk and Culture. University of California Press, Berkeley, CA.

Ehrlich, P. and G. Daily. 1993. Science and Management of Natural Resources. Ecological Applications **3**: 558-560.

Elster, J. 1992. Local Justice: How Institutions Allocate Scarce Goods and Necessary Burdens. Russell Sage Foundation, NY.

Fairhead, J. and M. Leach. 1995. False Forest History, Complicit Social Analysis: Rethinking Some West African Environmental Narratives. World Development, Vol. 23, No. 6: 1023-1035.

Fortmann, L. and E. Roe. 1993. On Really Existing Communities--Organic or Otherwise. Telos 95.

Friedman, L. 1984. Microeconomic Policy Analysis. McGraw-Hill Book Company, NY.

Friedman, T. 1994. I.M.F. Head Defends Russia Loan Policy Against Criticism The New York Times (National Edition), Vol. CXLIII, No. 49,595 (February 2).

Girard, R. 1987. Things Hidden Since the Foundation of the World. Stephen Bann and Michael Metteer, Translators). Stanford University Press, Stanford, CA.

Girard, R. 1989 (1986). The Scapegoat. Yvonne Freccero, Translator. Johns Hopkins University Press, Baltimore.

Gregory, W. 1989. The Defense Procurement Mess. Lexington Books.

Gross, J. and S. Rayner. 1985. Measuring Culture. Columbia University Press, NY.

Hardin, G. 1977 (1968). The Tragedy of the Commons. *In* G. Hardin and J Baden, editors. Managing the Commons. W.H. Freeman and Company. San Francisco.

Hilborn, R. and D. Ludwig. 1993. The Limits of Applied Ecological Research. Ecological Applications **3**: 550-552.

Hirschman, A. 1970. A Theory of Loyalty. *In* Exit, Voice and Loyalty. Harvard University Press, Cambridge.

Holling, C.S. 1993. Investing in Research for Sustainability. Ecological Applications **3**: 552-555.

Hukkinen, J., E. Roe and G. Rochlin. 1990. A Salt on the Land: A Narrative Analysis of the Controversy Over Irrigation-Related Salinity and Toxicity in California's San Joaquin Valley. Policy Sciences, Vol. 23, No. 4.

Lee, K. 1993. Greed, Scale Mismatch, and Learning. Ecological Applications **3**: 560-564.

Levin, S. 1993. Science and Sustainability. Ecological Applications **3**

Ludwig, D. 1993. Environmental Sustainability: Magic, Science, and Religion in Natural Resource Management. Ecological Applications **3**: 555-558.

Ludwig, D., R. Hilborn, and C. Walters. 1993. Uncertainty, Resource Exploitation, and Conservation: Lessons from History. Science **260**: 17, 36.

Luke, T. Forthcoming. Ecocritics: Contesting the Politics of Nature, Economy and Culture. University of Minnesota Press, Minneapolis, MN

Malinowski, B. 1954 [1948]. Magic, Science and Religion. Doubleday, Garden City, Y.

Mangel, M., R. Hofman, E. Norse, and J. Twiss, Jr. 1993. Sustainability and Ecological Research. Ecological Applications **3**: 573-575.

Marcuse, H. 1965. Repressive Tolerance. *In* Robert Paul Wolff, Barrington Moore, Jr, and Herbert Marcuse, A Critique of Pure Tolerance. Beacon Press, Boston, MA.

Mendeloff, J. 1988. The Dilemma of Toxic Substance Regulation: How Overregulation Leads to Underregulation at OSHA. MIT Press, Cambridge, MA.

Meyer, J. and G. Helfman 1993. The Ecological Basis of Sustainability. Ecological Applications **3**: 569-571.

Mooney, H. and O. Sala. 1993. Science and Sustainable Use. Ecological Applications 3: 564-566.

Moris, J. and J. Copestake. 1993. Qualitative Enquiry for Rural Development. Intermediate Technology Publications on behalf of the Overseas Development Institute, London.

Nichols, E. and A. Wildavsky. Forthcoming. Safer Power: Understanding and Improving Organizational Design, Human Performance, and Governmental Regulation in Nuclear Power Plants.

Orléan, A. 1988. Money and Mimetic Speculation. In P. Dumouchel, editor. Violence and Truth. Stanford University Press. Stanford, CA.

_____. 1989. Mimetic Contagion and Speculative Bubbles. Theory and Decision 27: 63-92.

_____. 1992a. The Origin of Money. In F. Varela and J-P Dupuy, editors. Understanding Origins. Kluwer Academic Publishers. Netherlands.

_____ (co-authored with Robert Boyer). 1992b. How do Conventions Evolve? Journal of Evolutionary Economics 2: 165-177.

Perrow, Charles (1984), Normal Accidents: Living With High Risk Technologies, Basic Books, New York.

Piccone, P. 1978. The Crisis of One-Dimensionality. Telos 35 (Spring).

Piccone, P. 1987/88. The Crisis of American Conservatism. Telos 74 (Winter).

Piccone, P. 1991. The Crisis of Liberalism and the Emergence of Federal Populism. Telos 89 (Fall).

Pitelka, L. and F. Pitelka. 1993. Environmental Decision Making: Multidimensional Dilemmas. Ecological Applications 3: 566-568.

Post, R. 1995. Constitutional Domains: Democracy, Community, Management. Harvard University Press, Cambridge, MA.

Roe, E. 1995. Except-Africa. World Development, Vol. 23, No. 6.

_____. 1994a. New Frameworks for a Very Old Tragedy of the Commons and an Aging Common Property Resource Management. Agriculture and Human Values, Volume XI, Number 1 (Winter).

_____. 1994b. Narrative Policy Analysis. Duke University Press. Durham, NC and London.

_____. 1994c. The Sharp Edge of the Sword. Transition 64.

_____. 1988. "Individualism versus Community in Africa? The Case of Botswana," Journal of Modern African Studies, Vol. 26, No. 2.

Rosenberg, A.A., M.J. Fogarty, M.P. Sissenwine, J.R. Beddington, and J.G. Shepard. 1993. Achieving Sustainable Use of Renewable Resources. Science, Vol. 262 (November 5).

Rothe J.P. 1991. Educating for Safety. Society, Vol. 28., No.3.

Rubenstein, D. 1993. Science and the Pursuit of a Sustainable World. Ecological Applications 3: 585-587.

Salwasser, H. 1993. Sustainability Needs More Than Better Science. Ecological Applications 3: 587-589.

Scharfstein, D. and J. Stein . 1990. Herd Behavior and Investment. The American Economic Review 80: 465-479.

Schelling, T. 1978. Thermostats, Lemons, and Other Families of Models. In Micromotives and Macrobehavior. W.W. Norton and Company, NY.

Scoones, I. 1994. Living With Uncertainty: New Directions for Pastoral Development in Africa. Overview Paper prepared for the Workshop on New Directions in African

Range Management and Policy, Woburn, UK, June 1993, International Institute for Environment and Development, London.

Shackle, G.L.S. 1972. Epistemics and Economics. Cambridge University Press. Cambridge.

Shiller. R. 1989. Stock Prices and Social Dynamics. Fashions, Fads, and Bubbles in Financial Markets. In Market Volatility. The MIT Press, Cambridge, MA

Shiller, R. and J. Pound. 1989. Survey Evidence on Diffusion of Interest and Information Among Investors. Journal of Economic Behavior and Organization 12: 47-66.

Sinfield, A. 1992. Faultlines: Cultural Materialism and the Politics of Dissident. University of California Press, Berkeley, CA.

Slobodkin, L.B. 1993. Scientific Goals Require Literal Empirical Assumptions. Ecological Applications 3: 571-573.

Socolow, R.H. 1993. Achieving Sustainable Devlopment That is Mindful of Human Imperfections. Ecological Applications 3: 581-583.

Sorensen, R. 1992. Thought Experiments. Oxford University Press, NY.

Thompson, M. 1993. The Meaning of Sustainable Development. Paper presented at a conference on "Global Governability," The Centre for Global Governance, The London School of Economics, 20th April.

Thompson, M., R. Ellis and A. Wildavsky. 1990. Cultural Theory. Westview Press, Boulder, CO.

Tiffen, M., M. Mortimore, and F. Gichuki. 1994. More People, Less Erosion: Environmental Recovery in Kenya. John Wiley, Chichester and New York.

Topol, R. 1991. Bubbles and Volatility of Stock Prices: Effect of Mimetic Contagion. The Economic Journal 101: 786-800.

United States Department of Agriculture, Forest Service; United States Department of Commerce, National Oceanic and Atmospheric Administration , National Marine Fisheries Service; United States Department of the Interior, Bureau of Land Management, the Fish and Wildlife Service, and the National Park Service; and the Environmental Protection Agency. 1993. Forest ecosystem management: an ecological, economic, and social assessment. Report of the forest ecosystem management assessment team ("FEMAT"). US Government Printing Office. Washington, DC.

Wheeler, S. 1995. Sustainable Urban Development: A Literature Review and Analysis, Urban Ecology, Inc. Oakland, CA.

Wildavsky, A. 1995. But Is It True? Harvard University Press, Cambridge, MA

Wildavsky, A. 1993. Accounting for the Environment (draft). To appear in the Journal of Accounting, Organizations and Society.

Wildavsky, A. 1988. Searching for Safety. Transaction Publishers. New Brunswick, NJ.

Wilson, J.Q. 1989. Bureaucracy: What Government Agencies Do and Why They Do It. Basic Books, NY.

Xenos, N. 1989. Scarcity and Modernity. Routledge, London and New York.

Zedler, J. 1993. Lessons on Preventing Overexploitation. Ecological Applications 3: 577-578.

PART III

A CASE OF SUSTAINABLE DEVELOPMENT

THE SHARP EDGE OF THE SWORD: A CASE OF SUSTAINABLE DEVELOPMENT

Whoever abandons the science of policy and behaves independently suffers from misery. Service to such an independent master is like licking the sharp edge of the sword.

From the <u>Sukra Nuti</u> (<u>Policy of Sukra</u>),[1] c. 800 AD

MAKING SUSTAINABLE DEVELOPMENT INTO A "CASE"

For many readers, the triangulated answers of Part II provide no straightforward policy and program for sustainable development or the science to underlie it. Readers will still want to know what to differentiate, what "needs" matter more, what to do about the unmanageable, and how to respond if population growth is not halted or percapita consumption reduced—in short, they want a kind of certainty that the complexity of the sustainable development issue does not at all provide. For others, it is enough to know that keeping options open for the future will always be complicated, never assured, and necessarily sensitive to all manner of local differences, contingencies, perverse feedback cycles, inevitable surprises, and case-by-case analyses that subvert the Cell 1 drive to generalize to all cases, everywhere, all the time.

Fortunately, Parts I and II of *Taking Complexity Seriously* set the stage for saying much more about sustainable development than that. Begin by noting how different Part II's triangulated answers are from most current recommendations for sustainable development. Many sustainable development advocates recommend, to quote Wheeler (1995, 15-16), "holistic and systemic thinking;" the triangulation instead recommends differentiating that thinking. Many advocates recommend "the need for a long-term perspective;" the triangulation instead recommends problematizing that need. Many advocates recommend "a search for a new balance between environment and economics;" the triangulation recommends instead searching for perverse feedback cycles that ultimately undermine the possibility of any such balance. And many

---

[1] de Bary (1958, 241).

sustainable development advocates recommend "a view of the planet as a whole;" the triangulation in contrast insists on first acknowledging just how unmanageable that planet is. These differences between conventional and triangulated recommendations reflect better than any other measure just how much of present thinking about sustainable development declares development to be complex, but then proceeds ahead making recommendations as if it weren't. Nor by this point should that conclusion be surprising, as the current views about sustainable development are largely a species of Cell 1 thinking literally orthogonal to the Cells 2 and 4 world in which our triangulation takes place.

As we have seen, the chief lesson of Part II is that to take the complexity of sustainable development seriously means treating "it"—keeping options open for the future—on a case-by-case basis. To put it another way: If there is a program or approach to sustainable development, it must be consistent with analyzing each case of sustainable development on its merits. To repeat: There can never be only one kind of sustainable development. Each is its own case. Since the notion of case-by-case analysis now moves centerstage, let us begin our search for further programmatic implications of the preceding analysis by asking just what it means to be a "case" of something. We do this, moreover, within the context of a case study on what has already been identified as sustainable development in practice.

One of the most popular forms of sustainable development on the ground has been "ecosystem management." Ecosystem management is described as the "use of an ecological approach in land management to sustain diverse, healthy, and productive ecosystems...[It] is applied at various scales to blend long-term societal and environmental values in a dynamic manner that may be adapted as more knowledge is gained through research and experience".[2] Fortunately, an example of ecosystem management—and our reformulation of such management in light of Parts I and II—allows us to illustrate what it means both to treat sustainable development on a case-by-case basis and to insist that sustainable development must itself be consistent with a case-by-case analysis of keeping options open for the future. Both points are developed in a case study of the controversy over the California Spotted Owl.

First, then, Just what do we mean by a "case"?

A SIMPLE, BUT POWERFUL, TYPOLOGY

As was introduced in Part I, begin by asking: Just what is an ecosystem *a case of*? Of what is an ecosystem an instance? What kind of case are we investigating when analyzing an ecosystem?

Charles Ragin (1992) develops a 2-by-2 typology for the question of "What is a case?" The two dimensions are (1) whether the "cases" you are analyzing are taken to be *empirical units* in the real world or *theoretical constructs* that are a consequence of the theories you hold or analytic frameworks you use, and (2)

---

[2] United States Department of Agriculture, Forest Service, and the United States Department of the Interior, Bureau of Land Management 1994.

whether these "cases," in turn, are taken to be *specific* in that they were developed during the course of your analysis or *general* in that they were pre-existing categories external to that analysis. (As we will see, a general case does not mean you can generalize across cases or beyond the case being analyzed.) Each of the resulting four types of cases (examples of which are in the California Spotted Owl controversy) can be seen at work in current understanding of ecosystems. I will be brief:[3]

(1) *Cases as specific empirical units*. Here, the ecosystem is to be found "out there" in reality. "In the first quadrant," according to Ragin, "researchers see cases [e.g., of ecosystems] as empirically real and bounded, but specific. They must be identified and established as cases in the course of the research process" (1992, 9).

(2) *Cases as general empirical units*. Here, the ecosystem is an already known, objective category used in the objective world. "In the second quadrant, researchers also view cases [of ecosystems] as empirically real and bounded, but feel no need to verify their existence or establish their empirical boundaries in the course of the research process...These researchers usually base their case designations on existing definitions present in the research literatures" (Ibid, 9-10). The assumption in many ecology textbooks, for instance, is that the ecosystems exist, whether or not one can clearly demarcate boundaries between them (e.g., Allen and Hoekstra 1992).

(3) *Cases as specific theoretical constructs*. Here, what you interpret an ecosystem to be depends on the interaction of the theories you hold and the research process you undertake. "Researchers in this quadrant see cases [of ecosystems] as specific theoretical constructs which coalesce in the course of the research. Neither empirical nor given, they are gradually imposed on empirical evidence as they take shape in the course of the research...Interaction between ideas and evidence results in a progressive refinement of the case conceived as a theoretical construct" (Ragin 1992, 10). This interactive refinement is at work in the process that has progressively led natural resource managers to see ecosystems as a potential planning unit for conserving habitat and biodiversity. (The period leading up to the Forest Service decision to adopt an ecosystem management approach is described much in these terms by Kessler 1993.)

(4) *Cases as general theoretical constructs*. Here, the ecosystem is the result of the combined disciplinary effort of natural scientists, which shapes and constrains the way they undertake their science and research. "Finally, in the fourth quadrant," writes Ragin, "researchers see cases as general theoretical constructs, but nevertheless view these constructions as the product of collective scholarly work. A researcher, for example, might conduct research on 'industrial societies' [or 'ecosystems'], recognizing that the assignment of

---

[3] While the focus is on ecosystems, the same four cases can be found to varying degrees in each of the close cognates of watersheds, habitats, biomes, and niches.

empirical cases to this theoretical category is problematic and that the theoretical category itself exists primarily because of collective scholarly interest" (1992, 10-11). This conception of ecosystem is most visible when members of the natural science community speak of ecosystem management as a "new paradigm" for natural resource management, one that in Kuhnian fashion displaces the older, more "linear" species-by-species approaches.

As we will see, this fourfold set of distinctions is crucial both to clarifying the limitations in many current views about ecosystem management and to reformulating ecosystem management so that it can work better in practice. Let us now turn to see how these different concepts of ecosystem surface in the controversy over the California Spotted Owl.

INTRODUCTION TO THE CONTROVERSY

Debate over the Northern Spotted Owl, particularly between logging and environmental interests, has dominated public attention in Washington, Oregon and northernmost California since the late 1970s. The controversy's national implications were showcased in President Clinton's 1993 Forest Summit. Northern California's own spotted owl began to move centerstage during the 1980s, as questions arose over the US Forest Service's strategy for managing the California subspecies.

In 1991, representatives from various federal and state agencies, along with other concerned parties, formed the Interagency Steering Committee for the California Spotted Owl to address these questions from the technical side. The Steering Committee established two advisory teams to assist it: (1) a Technical Assessment Team to make recommendations on the basis of the owl's biology, geographic distribution, habitat considerations, demography and the like; and (2) a Policy Implementation Team to suggest policies and guidelines for implementing the recommendations of the Technical Assessment Team. Each team submitted a report to the Steering Committee, and the following analysis is based on the two reports (hereafter the "Technical Report" [Verner et al. 1992] and "Policy Report" [Ruth and Standiford 1994]). Considerations of ecosystem management—and its accent on sustainability—figure prominently in each.

(1) *Technical Report* (1992). The Technical Assessment Team proposed a set of interim guidelines for maintaining spotted owls on California public timberlands. The guidelines develop several management areas and strategies for the spotted owl, and focus on factors such as age and size of tree, canopy cover, and fire management for maintaining the subspecies until more adequate regulations can be produced. These guidelines, which are set out in a main summary table, do not explicitly refer to ecosystems or ecosystem management. But their rationale, and more generally the Technical Report as a whole, do make these connections.

The Technical Report comes back again and again to issues of ecosystems and their management. The Report's first page (1992, v) declares:

> *For the forest ecosystems we consider, our appraisal must look many decades, even centuries, into the future to be certain that decisions we make today are sustainable into the future—that they do not compromise future benefits from these ecosystems.*

In this and similar passages, the Technical Report is asserting that ecosystems are out there, to be found in the external world and sustained there into the future. That is not to say ecosystems are static. All ecosystems are changing because of factors endogenous as well as exogenous to the ecosystem:

> *A fundamental assumption underlies management of owl habitat, as well as much of forest management in general: Ecosystems are inherently dynamic; they do not stand still. Changes take place both rapidly (through a variety of natural and man-made disturbances) and slowly (through climatic change and natural successional processes), and occur at many spatial and temporal scales. Changes have occurred in the past, and they will occur in the future even if we "do nothing."* (Technical Report 1992, 269)

As an example, ecosystems in California's Sierra Nevadas—the principal regional focus of the report—have been sculpted over the last 150 years by intensive sheep grazing, increasing fire suppression, and logging, among other factors (Ibid, 28).

Thus, while the Technical Report (Ibid, 30) speaks from time to time of "the natural conditions of forest ecosystems," overall it treats ecosystems as pre-existing cases to be found in reality and which have been shaped by human intervention over the longer term. Little, if any, time is given in the report to the sense that ecosystems are theoretical constructs of science or that each ecosystem is evolving into its own case of a highly localized fusion of theory to practice for meeting a highly localized objective.

The Technical Report does argue a kind of paradigm shift has been taking place in the move to ecosystem management. In the recent past, plans for national forestland in California were directed to achieving a "regulated forest" that could "control present [timber] harvest levels and plan future harvests to assure sustained yield within reasonable limits" (Ibid, 266). As the Report puts it in quoting another source, the shift from regulated forests to the "new paradigm" of managing ecosystems "involves a shift in management focus from sustaining yields of competing resource outputs to sustaining ecosystems" (Ibid, 267). One suspects that the repeated appeal in the Technical Report to ecosystems and their management stems from the authors' dissatisfaction with the conventional species-by-species approach they were working under when producing guidelines for spotted owl habitat.

(2) *Policy Report* (1994). Just as in the Technical Report, the Policy Report takes ecosystems to be out there in the real world to be managed. The latter's executive summary, for instance, speaks of "the natural role of fire...in regional ecosystems" and the need to "map habitats and ecosystems to assess their

fragility," all predicated on the assumption that there are ecosystems out there which are fired and surveyable (1994, 3 and 5).

Also as with the Technical Report, the Policy Report notes a paradigm shift is taking place in forest management:

> *Conservation and management of natural resources is [sic] in a state of transition. Evidence of the transformation to a more ecologically sensitive approach to management is reflected in the policies now being enunciated by the Forest Service...The Forest Service has embarked on an "Ecosystem Management"... initiative. Developments related to the Forest Conference held in the spring of 1993 in Portland [President Clinton's Forest Summit] echo these positions.* (Ibid, Chapter 7, 2)

Both the Technical and Policy Reports, in short, operate within a limited range of what ecosystems and their management are cases of. While each report argues that research should be iterative and dynamic (more below), each little recognizes that such learning will inevitably widen the range of what ecosystems and their management are taken to be cases of.

In the terminology of Part I, none of this would matter if the reports' actual use of the ecosystem convention stabilized the assumptions for ecosystem management in the face of high uncertainty and complexity. Sadly, there are problems here. The first is that we do not know much about ecosystems even within existing management contexts, let alone future ones:

> *No certainty exists that ecosystem processes, whether the population dynamics of spotted owls or successional trends in multi-species plant communities, will be maintained in regulated forests. We propose that the [Forest Service] move away from its goal of regulating forests and develop, instead, the knowledge to manage for more natural stand structures, compositions and functions. Given incomplete knowledge of natural forest conditions and, once known, the uncertainties of how to produce such stands, this is clearly an area for high-priority research.* (Technical Report 1992, 30)

Similar considerations hold for the wider context in which ecosystems operate: "Economic and social systems have an inherent degree of unpredictability," in words of the Policy Report that echo the *Ecological Applications* debate. Accordingly, "developing mitigation measures and integrating them into policy implementation must be a dynamic, iterative process" (1994, Chapter 5, 49).

Another problem is that what little we do know about ecosystems indicates these processes are very complex. "Complex" is a term that occurs repeatedly in the Technical Report (1992, 172, 261, passim). The landscape used by the owl is often "complex and heterogeneous", fire has left a "complex...pattern", and the landscape itself is but only one of a complex of land categories relevant to the spotted owl management (habitat type, home-range, stand, and territory, as well as, more administratively, provinces, and public/private land-use categories). Figure 4G of the Technical Report best pictures the complexity by showing a maze of intersecting arrows representing over twenty important causal factors said to have an impact on the spotted owl, with the disarming caption "*simplified, schematic* representation of *some* important ecological linkages..." (Ibid, 71; my italics).

The complexity has definite implications for ecosystem management in the Technical Report's view. "Understanding a species' ecology and life history and its role in ecosystem functions is vital to successful management of that species" (Ibid, 55). Because ecosystems are complex, so too must be their management: "It will be more complex and expensive," according to the Report (Ibid, 271). For example, "[a] great deal of sophisticated silviculture will be required" (Ibid, 267). As the quotes indicate, even if ecosystems were simple, their management would still be complex, if merely because of the uncertainties over how to achieve the objectives pursued by the management. In particular, a primary management goal appears to be sustaining an ecosystem that has yet to re-materialize, namely, "a pattern of stand structures that mimic historical stand conditions" (Op Cit). As the Report puts it, "we need to begin changing the forest structure back to a form more akin to historical patterns" (Ibid, 273). That this presettlement pattern involves "complex stand structures" simply reinforces the complexity of the practices needed in ecosystem management.

Both the Technical and Policy Reports recommend "adaptive management" as the way to respond to these complexities, and it is this adaptive management which is at the heart of their conception of what ecosystem management should be. Under the heading "An Adaptive Approach to Mitigation," the Policy Report concludes:

> *Designing an approach to implement mitigation [of reduced timber harvesting to safeguard the spotted owl] must be based not only on static analysis of impacts but also on the dynamics of the policy...As with the management of natural processes, management of social and economic processes requires careful monitoring and a "feedback" mechanism that uses this information to improve future mitigation measures and ease the transition.*
> (1994, Chapter 5, 49)

The authors of Technical Report make it quite clear that learning how to manage an ecosystem will necessarily be an incremental and iterative process of experimentation, re-experimentation, continuous hypothesis-testing, feedback, and gradual trial and error, in short "adaptive management." It is the "process of implementing policy decisions as scientifically driven management experiments that test predictions and assumptions in management plans, and using the resulting information to improve the plans".[4]

---

[4] United States Department of Agriculture, Forest Service; United States Department of Commerce, National Oceanic and Atmospheric Administration , National Marine Fisheries Service; United States Department of the Interior, Bureau of Land Management, the Fish and Wildlife Service, and the National Park Service; and the Environmental Protection Agency 1993.

Unfortunately, there is little acknowledgment in either report that such management will leave behind as many, if not more, research and management uncertainties as it resolves. I find only one reference where the Technical Report (1992, v) recognizes that "the way of science...inevitably leaves us with areas of uncertainty" (for an early discussion of this frequently remarked-upon finding, see Cohen and Weiss 1977). Elsewhere in both reports the assumption is that this kind of research-oriented management will reduce uncertainties and get us to an ecosystem management that sustains the kinds of ecosystems desired. Even if uncertainties were actually reduced *on net* through adaptive management, we would still face the distinct possibility that the management we end up with may be much more interactive than it is incremental. As our thinking about "What is an ecosystem a case of?" evolves, there will be times when we realize that the uncertainties we are reducing are simply not the uncertainties we should have been worrying about in the first place.

Once again, then, we are back to the Ragin typology.

THE IMPLICATIONS FOR ECOSYSTEM MANAGEMENT AND A CASE-BY-CASE SUSTAINABLE DEVELOPMENT

Ecosystems and their management are much richer than the Technical and Policy Reports imply. Below are ten suggestions for ecosystem management that follow directly from the Ragin typology of ecosystem concepts. In the process, Parts I and II are drawn upon to show the programmatic and policy implications of the preceding analysis, namely, how ecosystem management reformulated in light of these suggestions is an example of sustainable development as a case-by-case approach to keeping resource options open for the future.

Inevitable Differences

First, we should not be surprised—nor, for that matter, even bothered—by the fact that very different conceptions of "ecosystem" are at work when people seek to manage ecosystems. No one of the four ideal "cases of ecosystems" is *a priori* more useful than the others. More to the point, resource managers and scientists cannot be expected to privilege one type only. There will always be textbooks, shifting paradigms and interactive research processes—and unavoidably so. Or, in the terminology of Part II, the injunction, *Differentiate!*, has to be applied to "ecosystems" before one can even think of their "management."

Inevitable Confusion

Thus, it is wrong to believe that you are advancing knowledge simply by finding that, e.g., range ecosystems are more complex in the field than in a textbook, such as Stoddart, Smith and Box (1975). All you are doing is confusing different concepts of range ecosystems.[5] Nonetheless, some confusion is inevitable, and it must be expected that resource managers and scientists will from time to time talk at cross-purposes around the same table.

The Red Herring of Boundaries

Several issues become easier to address when we recognize that differing answers are likely to the question, What is an ecosystem a case of? First, the notion that ecosystem management somehow depends on specifying the boundaries of ecosystems falls away. Resource managers who believe that ecosystems are given or objective or interactive or collectively-agreed upon are unlikely to jettison the ecosystem concept simply because the perimeter of any such system is ill-delineated. What matters more than boundary is the set of biological, physical and ecological interactions within the ecosystem which people agree to manage as part of "ecosystem management." There are thus multiple reasons why arbitrary (i.e., case-specific) decisions and compromise—over the perimeter of the managed area and the set of interactions to be managed therein—are to be expected and indeed invited as part of the "management" side of ecosystem management.

The Necessary Evolution of Ecosystem Management

If ecosystem management is necessarily a constellation of different concepts and conventions over just what an ecosystem is a case of, then it also comes as no surprise that management is more likely to be effective when it legitimates these differing understandings rather than obscures, avoids, or forces us to choose among them. One such way to legitimate the co-existence of multiple cases of ecosystems is to insist that ecosystem management is evolutionary: Managers start with the expectation that the ecosystem is out there waiting to be identified, realize once in the field that there are problems in delineating system features, later acknowledge that such problems arise in part because what is out there depends crucially on how what "it" is they are looking for is defined in the first place, and then end up in better understanding that what works best in any particular situation is a function of tailoring theory and practice to meet the specific objectives agreed upon in ecosystem

---

[5] Note these differing cases would remain even if resource managers and scientists did not entertain different theories. The problem is not that some experts subscribe to, e.g., range succession theory and others to state-and-transition models. Even if everyone subscribed to the same framework, the cases of ecosystems would still differ *as long as* the subscribers did not agree how to balance the Ragin typology's dimensions of empirics and theory and the general and specific in the understanding and management of ecosystems.

management, case-by-case.[6] Thus, while ecosystem management may rarely start out case-specific, it will have to end up so in practice, if it is to respond effectively to locally varying circumstances. This conclusion should come as no surprise in light of the Part II findings.

The Argument So Far

To summarize, if you accept the typology of "cases of ecosystem," then you also should accept that (i) there is no single conception of "ecosystem," so don't worry about it (even moreso as there are other equally valid conceptions); (ii) there is no single way to bound the perimeter of ecosystems, so start with the recognition that boundaries are always case-specific (in short, there may be nothing incongruous about managing ecosystems within pre-existing administrative areas); and (iii) since the real issue is one of agreeing over what set of interactions within the ecosystem are to be managed in order to have a case of "ecosystem management," the place to start is by asking, What interactions, if any, are common across all cases of ecosystems?

Abilities Matter Most

As for the choice of common interactions for ecosystem management, it depends not on the science or needs of ecosystems, but on what sets of interactions resource managers and scientists are *able* to manage. What they are able to manage, in turn, depends fundamentally on what they take ecosystems to be cases of. Assume, *contra* the Technical and Policy Reports, that ecosystems, in addition to being objective units, are at other times constructs of the scientist's and manager's mind or the interactive product of science and management or the moving target of a science and management that are themselves evolving. You end up having to rely on approaches that are vastly more varied than the computer and GIS technologies recommended in the two reports. At a minimum, the approaches include an ecosystem management team whose members are comfortable with treating different cases

---

[6] The way cases of ecosystems are articulated in practice may, of course, differ from the terminology of the Ragin typology. For example, there are three very different models of "rangeland development" discussed today internationally. Model I is be the preservationist's ideal—a totally fenced-off range ecosystem, left to its own dynamics, forever untouched by human hands, even the herders. Model II would seek to mimic the imagined past, e.g., a restored grasslands inhabited by many wildlife, few cattle and even fewer herders, all living as they did before. Model III would be the development of sustainable communities of herders in more sustainable relationships to the rangelands and natural resources around them. The descriptions have an element of caricature, but each model does represent a progressively more interactive notion of a range ecosystem consistent with the Ragin typology. From found and untouched, through sculpted and maintained, to evolving and unstable, they represent very different cases of ecosystems.

of ecosystems as legitimate. One way to ensure this is to have team members work at scales different *both* in kind *and* in degree, e.g., the social scientist works on a sector, the hydrologist over a landscape, the plant ecologist in a niche, and residents within a community. (Adaptive management and ecosystem management are multiscale, but the scales constitute nested hierarchies different only in degree, e.g., site, watershed, river basin and region.) An ecosystem management that focuses more on what it is able to manage than on what needs to be managed follows as well from the findings of Parts I and II. Indeed, one of the best ways to observe the injunction, *Problematize need!*, is to insist abilities differentiate behavior more effectively than do needs in some cases.

Multidisciplinarity Works Best When It Is Case-By-Case

Another way to put the preceding point is, Avoid thinking that a team is multidisciplinary simply because it is a mix of professionals, e.g., a forest ecologist, a forest products economist and a population biologist. In the latter case, the professions are different, but their operating scales may well be too similar for the purposes of establishing different cases of ecosystems. One considerable benefit of having a multidisciplinary team organized around members who work at different scales and according to different ecosystem concepts is the increased likelihood that (i) each team member will have to rely on his or her own best judgment about specific interconnections between scales and (ii) the team as a whole will thereby be better able to treat each case of ecosystem management on its merits. To repeat, one major way to respond to ecosystem complexity is to insist that their management has to be done case-by-case basis, with management strategies differentiated in important respects from ecosystem to ecosystem.

In Ecosystem Management, Top-Down Versus Bottom-Up
Planning Doesn't Matter as Much as Outside-In Versus Inside-Out
Planning

The most pressing problem with any multidisciplinary planning for ecosystem management is not that the planning by experts is top-down rather than bottom-up, but that it is most likely *outside-in planning*. Assume the planning requirement for ecosystem management is that all the stakeholders "in" the ecosystem are at the table to hammer out a management plan for the area in question (indeed a requirement of some legislation). Assume the stakeholders are developers, environmentalists, local leaders, and state government officials and experts. What if the developer owns land in the area but actually lives and works elsewhere, the environmentalist visits the area only on weekends, and state officials come from the capital to the area only for meetings? They may be *de jure* stakeholders, but do they have the same *de facto* "stake in" the ecosystem as do local leaders and residents? Are all "stakes" equal? Today stakeholder planning isn't so much bottom-up in contrast to top-down as much as it is an outside-in planning that, in the terminology of Part II's Critical theory, homogenizes outsiders and insiders, experts and residents, into equivalent "stakeholders," simply by asserting that the claims of expertise or a public trust interest in the ecosystem is equivalent to the claims of residence there.

The Way Inside-Out Planning Should Work

The obvious challenge is to come up with varieties of *inside-out planning* for ecosystem management, where local leaders and residents are themselves the experts and where the planning process is itself initiated and guided from within the local ecosystem. The developer, environmentalist and state official in our example would not be excluded, but they come to the table only as outsiders living and working in other areas, e.g., via locally-initiated teleconferencing and interactive electronic mail with local leaders and residents. Why is inside-out planning better than outside-in? Because, as we saw in Part I, under conditions of high complexity, information becomes much more policy-relevant when those who gather it are as well those who end up using it—and this is much more probable under inside-out planning than otherwise. To put it another way, the virtue of inside-out planning is that it has a better chance of creating Cell 1 situations where causal processes and understandings are clearer. If, however, we are stuck with a planning process where any interested party is a potential participant to in the planning (e.g., by virtue of the public trust doctrine), then it is all the more imperative to ensure the process is evolutionary and one that legitimates there being different (case-specific) answers to "What is an ecosystem a case of?"

Why Process Matters

What if none of this works? What if ecosystems remain unmanageably complex and their interactions intractably uncertain, no matter how sophisticated our technologies or simple our management strategies? The answer is not sweet: The best we can do under these circumstances is to commit ourselves to a *process* whereby specific ecosystem definitions fall away, are replaced by new ones, and evolve through time. It is this process which ends up stabilizing the assumptions for management, even when the concept or convention of the moment may becoming increasingly unworkable. One such process has already been put into place through the Clinton administration's effort to fund and develop Adaptive Management Areas. As strategies develop locally in these areas, there will be (and already has been) diffusion and interchange between areas and later between localities that adopt and adapt these strategies to meet their own local circumstances. To adapt a point made in Part II, perhaps the very best we can hope for under conditions of high uncertainty and complexity—ecological and societal—is to have ecosystem management (i.e., one kind of sustainable development) practiced somewhere and all the time, though not everywhere and always in the same place.

FINAL REMARKS: SCIENCE, POWER, AND POLITICS

The Technical and Policy Reports imply that politics, not complexity, is the real obstacle to realizing ecosystem management. So too you hear for sustainable development: Yes, the world is complex, but...politics are to blame...government bureaucracy is killing us...corporations rule the world... If you really believe the world you operate in is fundamentally complex in a Cells 2 and 4 sense, then the Cell 1 argumentation of the preceding sentence is simply not plausible—except when a case-by-case analysis finds instances where the policy issue of interest in your world abides by a truly tightly coupled and linear process.

Yet if the Technical and Policy Reports are any guide, natural scientists and related experts are all-too-ready to adopt a kind of Cell 1 thinking that insists on an *a priori* separation of politics and science in the face of high uncertainty and complexity. According to the Policy Report (1994), experimental research sponsored through adaptive management should be evaluated on its success in advancing knowledge, not in mitigating the more politically-charged economic impacts of reduced timber cuts in order to save the owl. The Technical Report (1992, p. 33) goes further in seeking to separate politics from science: "The primary barrier to finishing what is needed to remove the remaining uncertainty about the status of the California spotted owl is," in its view, "administrative, not scientific." If these barriers cannot be removed, "then the political will is missing" to undertake the required management for sustaining spotted owls specifically and ecosystems generally. In so arguing, the Technical Report comes parlously close to using politics as an excuse not to think through just how *any* complex ecosystem management strategy can ever work, let alone succeed, if ecosystem processes and goals to be optimized are as complex and uncertain as the Technical Report itself identifies.

The crux of the matter is the Technical Report's notion of "political will." At the most palpable level, the real problem is less the *lack* of political will as the *excess* of it: As a group, politicians and agencies in this country are committed to everything, and *that* is a problem. But there is a subtler, more important issue. In situations of high complexity and uncertainty, "political will" should be seen not in the direct power terms of the Technical and Policy Reports, but rather as an infrequent "policy window".[7] In the terminology of Kingdon (1984), a policy window is created by the unpredictable tangency of three streams that wend their way through a complex world of loose coupling interspersed with moments of tight coupling. The three streams—problems, policies and politics—develop

---

[7] Remember, power was a principle notion driving the *Ecological Applications* debate, at least from the perspective of Ludwig, Hilborn and Walter's statement (1993, 547), i.e., "[w]ealth or the prospect of wealth generates political and social power that is used to promote unlimited exploitation of resources."

> *...and operate largely independently of one another.
> Solutions are developed whether or not they respond to a
> problem. The political stream may change suddenly whether or
> not the policy community is ready or the problems facing the
> country have changed. The economy may go sour, affecting the
> budget constraint, which imposes a burden on both politicians and
> policy specialists that was not of their own making. The streams
> are not absolutely independent, however. The criteria for
> selecting ideas in the policy stream, for instance, are affected by
> specialists' anticipation of what the political or budgetary
> constraints might be...Despite these hints of connections, the
> streams still are largely separate from one another, largely
> governed by different forces, different considerations and
> different styles....Once we understand these streams taken
> separately, the key to understanding agenda and policy change is
> their coupling. The separate streams come together at critical
> times. A problem is recognized, a solution available, the political
> climate makes the time right for change, and the constraints do not
> prohibit action. Advocates develop their proposals and then wait
> for problems to come along to which they can attach their
> solutions, or for a development in the political stream like a
> change of administration that makes their proposals more likely to
> be adopted...I label an opportunity for pushing one's proposals a
> "policy window"...(Kingdon 1984, 93-94)*

In these terms, power—and this includes political will—comes about in those uncommon, complex moments when policies, problems and politics converge so that the problems of the moment are tangent to the politics of the moment which in turn are tangent to the policies of the moment. In these *cases*, multinationals do have power, the Forest Service does have political will (as was the case when Jack Ward Thomas became the head of the Forest Service and championed ecosystem management). But such political will is truly, truly a matter of contingency. It could well have been otherwise and will be otherwise, when the streams meander off in their own directions and with great surprise, *as they inevitably do in Cells 2 and 4 situations*. To say that something now has "power" in a Cell 2 and 4 world is simply to say the present is an alternate version of what could have happened instead. Here, complexity creates power.

It borders then on the tautological to note that you cannot make policy windows happen, you can only take advantage of them when they unexpectedly arise. Ecosystem management may or may not have been just such a policy window. Whatever the case is (and we know there will be different cases of ecosystem management to evaluate), ecologists and environmental scientists do a great disservice when implying, as they do in the Reports, that it is politics, and not the complexity itself, which is the real impediment to realizing ecosystem management. To revert again to Kingdon, the stream of politics isn't the problem; rather it is the contingency and complexity of the three streams of politics, problems and policies that make policy windows so few and far between.

Thus, when ecologists insist that ecosystems are really too complex to be managed, what they are really promoting is an ecology that keeps the politics, problems and policies associated with ecosystem management separate and apart. In such circumstances, policy windows become even more remote, the chances for successful ecosystem management dwindle accordingly, and the politics of ecosystem management keeps flowing along on its own, without intersecting science-based problems and policies. We simply cannot afford this kind of ecology. We cannot afford a science that is addressing what we are unable to manage, when there are things we should and could be managing instead. An ecology that fails to keep management simple is a science that has truly failed to address complexity.

If I have been harsh on scientists, let me now be harsher on the politics. Allow me to conclude *Taking Complexity Seriously* not on the stream of politics we have, but on the more difficult politics to which this book's analysis leads us, be it for ecosystem management specifically or sustainable development generally. The classic method for stabilizing any system—biological, engineering, organizational, socioeconomic—in a complexly turbulent environment is to decouple that system from that environment as best as one can. What then would such a "politics of complexity" look like if it was organized around such decoupling? There are no firm answers here, but available literature, much of it discussed in Parts I and II, triangulates on the following (see also Roe 1994a, b). It is fitting that a book so dedicated to triangulation should end with one.

First, a politics of complexity would remind us that loose coupling itself is rational in a complex world, where system delays, reserves and flexibility are not only required but desirable. Such a politics would recommend decentralization if you are worried about coping with surprises in situations where delay and unpredictability are endemic. It implies that what works better is often a smaller scale of decisionmaking, where information gatherers are, again, the users of that information. It suggests that the fewer steps in implementation and the more interdependent the probabilities (where success at one step enhances the chances of success at the next step), the greater the chances for implementation as planned (precisely what is found in a number of localities and communities). It pulls us to the local level where all those Cell 1 notions we were brought up to believe in—power politics, learning curves, getting the "big picture"—have a better chance of being matched by Cell 1 situations in which these notions are appropriate. It maintains that some processes and systems are cases in their own right and have to be analyzed as such. It argues fiercely against top-down or outside-in planning and for community-based, inside-out management. It notes there are some important allocations systems—e.g., Elster's local justice systems—which really are local and localized, that is, semi-decomposable from larger systems and must be treated as such. It resists a global analytic that scales up, rather than down, policy issues more effectively dealt with locally or regionally. It insists on distinguishing between insiders and outsiders to any resource management system. It recognizes that no matter how interconnected the world is causally, people's stories about this world—their arguments and scenarios—are typically much less connected. Finally, it accepts the fact that sometimes the most useful analysis you can undertake is to use your own best judgment as to what are the merits of the case. This interdigitated landscape of loose coupling, decentralization, smaller scales where information gatherers and users can be the same, locally interdependent probabilities, the local level as Cell 1 situations, cases in their own right, communities, local justice, semi-decomposable subsystems, scaling down, insiders and outsiders, and judging the merits of each case is the landscape, I believe, disclosed to us by a politics of complexity. Moreover, it is a landscape to which all of us are being driven, whether we like it or not. As the world becomes more complex and turbulent, one can expect more and more pressure to decouple from that world, particularly when that is the only way we have left to stabilize the assumptions for decisionmaking in the face of a complexity and uncertainty that is out of our control.

What does this mean practically? It suggests the increasing complexity of the global economy will give rise to the increasing fissiparousness of secessionism (not just in the former Yugoslavia or USSR, but in the US as well), whatever one may feel about the good or bad of secessionists, nationalists and nativists. As critics are quick to remind us, there are many countries in the world that you can't talk about without addressing the effects of chaotic urbanization, unemployment, economic recession, generational conflict, the legacy of ethnic strife, and attendant crises. All that, and more, is true for these countries. Is it any wonder, then, that when confronted with expedient leaders, wretched public management, runaway crime and economic chaos, many of their citizens want to detach themselves from their nation as presently constituted? Is there any doubt the rest of us would respond similarly? For these reasons alone, secessionist pressures to decouple can be expected to intensify as the problems of that world become more numerous, various, and interrelated—in a word, complex. An increasingly Cell 2 and 4 world both pushes and pulls us to create more Cell 1 situations. While globalization and localization are often taken to be odds (this is why rapid change globally can make things so turbulent locally), the former is quite clearly accelerating the latter.

Indeed, secessionism and the case-by-case approach this book has espoused for sustainable development are but two sides of the same coin of complexity. *Indeed, I can't imagine a more imperative political issue than how the new entities to be thrown ashore from the wave of secessionism ahead—many smaller, some larger than current nation states—are to become sustainable.* I hope that in showing how sustainable development is really possible only on a case-by-case basis (say, within a framework of inside-out planning), I have gone some way in convincing the reader that the inevitable secessionist (nationalist, religious, ethnic) movements coming down the pike must as well be evaluated on precisely the same basis. No blanket condemnation of nationalists, populists and nativists will do anymore. Such Cell 1 thinking will only mean that we once more risk putting tongue to sword, with nothing left but inspiration to guide us from there.

REFERENCES

Allen, T. and T. Hoekstra. 1992. Toward a unified ecology. Columbia University Press. New York, NY.

Cohen, D. and J. Weiss. 1977. Social Science and social policy: schools and race, in C. Weiss, editor. Using social research in public policy making. Lexington Books. Toronto.

de Bary, Wm. Theodore, editor (1958), *Sources of Indian Tradition*, Vol. 1 (A. Basham, R. Dandekar, P. Hardy, V. Raghavan, and R. Weiler, compilers), Columbia University Press, New York.

Kessler, W. 1993. Evolution of range ecology practices and policy: back to our rangeland roots. Rangelands **15**: 101-103.

Kingdon, J.W. (1984), *Agendas, Alternatives, and Public Policies* , Little/Brown, Boston.

Ragin, C. 1992. Introduction: cases of 'What is a case?' *In* C. Ragin and H. Becker, editors. What is a case: exploring the foundations of social inquiry. Cambridge University Press. Cambridge.

Roe, Emery (1994), *Narrative Policy Analysis*, Duke University Press, Durham, NC.

Roe, Emery (1994a), "Against Power, For The Politics of Complexity," in *Transition* 62.

Roe, Emery (1994b), "The Sharp Edge of the Sword: Reply to My Critics," in a special symposium on my article "Against Power, For the Politics of Complexity," in *Transition* 64.

Ruth, L. and R. Standiford (compilers). 1994. Conserving the California Spotted Owl: Impacts of Interim Policies and Implications for the Long-Term. Report 33. Wildland Resources Center, University of California. Berkeley, CA.

Stoddart, L., A. Smith, and T. Box. 1975. Range Management. McGraw-Hill. New York.

United States Department of Agriculture, Forest Service, and the United States Department of the Interior, Bureau of Land Management. 1994. Final supplemental environmental impact statement on management of habitat for late successional and old-growth forest related species within the range of the Northern Spotted Owl. Volume I (Glossary). Portland, OR.

United States Department of Agriculture, Forest Service; United States Department of Commerce, National Oceanic and Atmospheric Administration , National Marine Fisheries Service; United States Department of the Interior, Bureau of Land Management, the Fish and Wildlife Service, and the National Park Service; and the Environmental Protection Agency. 1993. Forest ecosystem management: an ecological, economic, and social assessment. Report of the forest ecosystem management assessment team ("FEMAT"). US Government Printing Office. Washington, DC.

Verner, J., K. McKelvey, B. Noon, R. Gutierrez, G. Gould, Jr., T. Beck. 1992. The California Spotted Owl: a technical assessment of its current status. General technical report. PSW-GTR-133. Pacific Southwest Research Station, Forest Service, U.S. Department of Agriculture. Albany, CA.

INDEX